SEASHELLS
of the Pacific Northwest

By James Seeley White

Both the sheltered waterways of British Columbia and Washington and the rugged coastline of Oregon and Washington abound in interesting shells. Often these natives of the Pacific Northwest have been overlooked in favor of exotic shells imported from warmer seas. *Seashells of the Pacific Northwest,* covering eighty-three of the most available shells, is designed for the average beachcomber, skin diver, and student who loves the sea but knows little about the treasures found there. Readers may be surprised to learn of the three-, four-, and five-inch shells found in Northwest waters, and of the bright colors that many shells display.

The author shares his observations of the animals' behavior, his secrets for finding them, and tips for preserving them. Each illustration identifies where the shells were found, and a special section on photography explains how the pictures were taken.

Seashells of the Pacific Northwest uses simple, explanatory language geared for the beginning shell enthusiast. It discusses those shells normally encountered through careful observation during walks by the sea, and adventures beneath it. The material prepares the reader for further study, should this be sought, and gives collectors sufficient information for differentiating between related species.

The book is intended for light, informative reading. The author hopes that the reader, through these pages, will share some of his pleasure in observing and collecting the seashells of the Pacific Northwest.

SEA

of the

By James Seeley White

with technical editing
By C. Dale Snow

SHELLS

Pacific Northwest

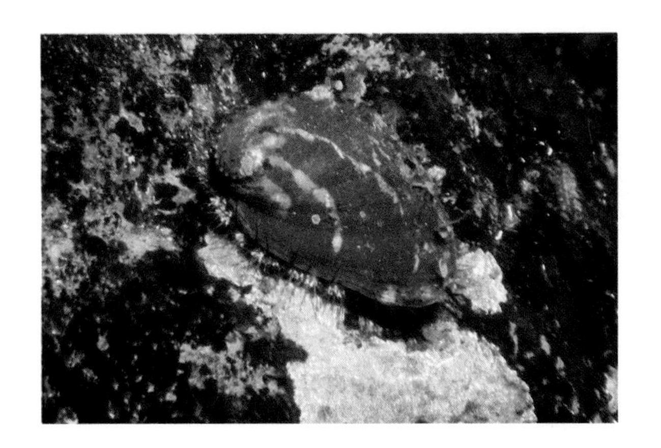

Binford & Mort

Thomas Binford, Publisher

2536 S.E. Eleventh • Portland, Oregon 97202

Seashells of the Pacific Northwest

First Edition
1976

Library of Congress Cataloging in Publication Data

White, James Seeley.
Seashells of the Pacific Northwest.

Includes indexes.
1. Shells—Northwest, Pacific—Indentification.
I. Title.
QL417.W48 594'.04'7 73-89239

ISBN 0-8323-0232-5
ISBN 0-8323-0233-3 pbk.

CONTENTS

PREFACE

This book is not intended as a complete scientific treatise on the shells of the Pacific Northwest. Its purpose, instead, is to introduce the reader to our principal seashell families and to provide information about the animals that produce the shells—how they live and where to find them. Thus, we have included the Northwest seashells most apt to be noticed by a beachcomber or a skindiver.

The shells described, all from animals in phylum Mollusca, are grouped in the separate classes of Gastropoda, Polyplacophora, and Pelecypoda. Within these classes we have omitted the level of order and gone directly to the more clearly meaningful family level of classification. A general description of the worldwide families into which Northwest shells fit is first given in the box which precedes the individual Northwest shells. Shells within each family are identified by one common name (though each may have several) and, in italics below the common name, is the internationally accepted scientific name. The scientific generic name (identifying the genus to which the shell belongs) is shown first and is capitalized; the species name appears second and is not capitalized. Following the scientific name, though not in italics, is the name of the author or scientist who first described the shell and published the name. If subsequent studies have transferred the species to another genus, the name of the original author is placed in parentheses.

Serious shell collectors will notice that several generic names and some species have changed from those with which they are acquainted. These changes represent the most recent updating of scientific names of which the technical editor is aware. The most noteworthy change is that the genus *Acmaea* has been replaced by *Collisella* or *Notoacmaea* in most cases. The need for such a revision has long been recognized by invertebrate zoologists. The Purple-hinged Rock Scallop has been returned to the specific name *giganteus.* Though a little less descriptive than *multirugosus*, it does predate the latter by over 100 years and should be retained.

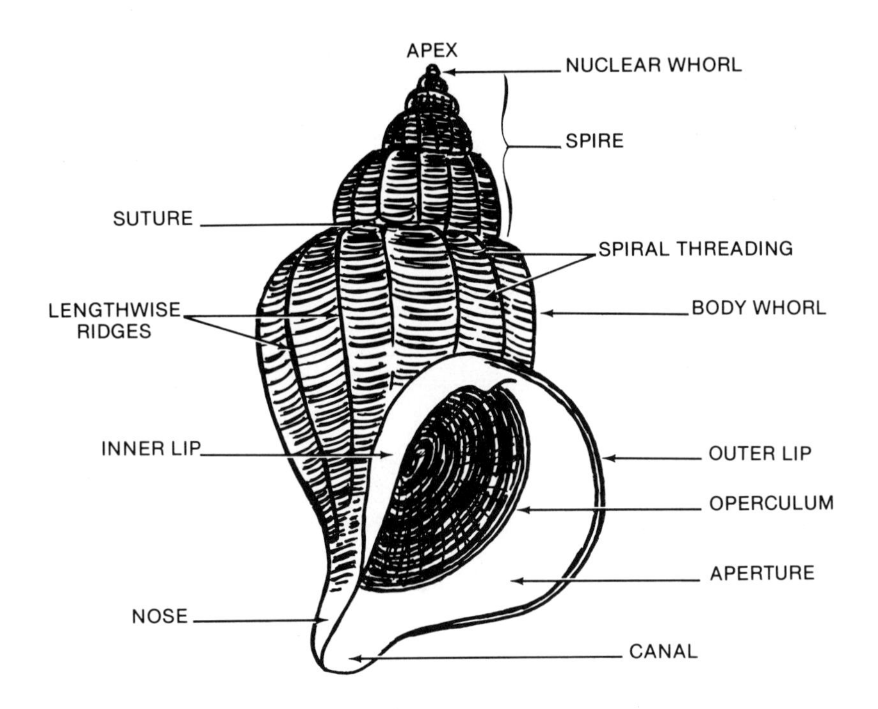

Diagram of Snail

PART ONE

SNAILS

Shells of marine snails have been collected since ancient times. Like clams and oysters, the animals that build them belong to phylum Mollusca (Mollusks), a division of the animal kingdom which includes more than 60,000 species. Because snails are soft-bodied and slow-moving little creatures, shells are their best protection, and these appear in many fascinating forms and colors—from the long-spired bittiums to the flat, wide-shelled abalones. Some, like the limpets, have conical, cup-like shells; others, like the moon snails, are rounded and coiled. Inhabitants of quiet waters (either sheltered coves or waveless depths) may develop thin shells, but those of the storm-battered coast must build thick ones for protection against the pounding surf. However they have developed, their single shell designates them as univalves, in contrast to the double-shelled clams, oysters, and scallops, which are bivalves.

Snails are also known as gastropods (*gastro* or stomach + *pod* or foot) because of their strange method of locomotion. A large foot, muscular and flat, comprises the ventral or stomach surface of the animal and is used for creeping, digging, holding prey, and prying apart the shells of clams to eat the contents.

The physiology of various snails also affects the shape of their shells. Olive and basket snails travel under the surface of the sand and have a short, upturned canal to house the siphon, which draws oxygen-laden water from above. Predator snails usually have a longer canal, while those that scrape algae from rocks—like limpets, turbans, and abalones—have no canal because they breathe differently.

Rock Dwellers

Rocky intertidal area, Lincoln City, Oregon.

Rocky intertidal and shallow-water areas of the Pacific Northwest coastline are ideal for collecting shells or studying marine life. These areas are on the fringe of the greatest population density of the sea—the range of light penetration. Along the water's edge grow little garden plots, sometimes acreages, of the vegetables of the sea, the seaweeds or marine algae. Around and among these marine plants live many snails that browse upon them, such as turbans, limpets, and abalones. In places not covered with seaweed, colonies of mussels and barnacles thrive, providing food for the predatory dogwinkles, whelks, and hornmouths. Some snails here are so dependent upon rocky areas that they are unable to move about on a surface not firm like a rock. At times they cannot even right themselves if torn from their rocky home.

To these intertidal areas come shell collectors, students of marine life, and curious amateurs, for here they can see great numbers of snails and other marine animals. During periods of exceptionally low tides they may catch a glimpse into the lives of these animals, perhaps even take away a few colorful shells.

A cluster of BLACK TURBAN snails huddle in a crevice at Cape Arago, Oregon, to await the return of the sea.

Entering a rocky intertidal area while the tide is out is something like entering a darkened theater—one's eyes must become accustomed before all of the action is apparent. Then, after this period of adjustment, one can find what the creatures of the ocean do when the sea is temporarily absent. Some may be found hiding among the clumps of seaweed, waiting while staying cool and moist. Many scramble under rocks or into crevices for protection. A few, like the turbans and limpets, are so well adapted to this changing environment that they can simply clamp down tightly, sealing in their own little corner of the world.

Knowing this, we learn that turbans and limpets may be found in sheltered areas of rock, dogwinkles in crevices and the under sides of boulders, and other snails may be hidden in the seaweed. There is, however, a segment of this area where the more vital life of the ocean still goes on. This is the tidepool area —the place the most sensitive creatures seek, where fish still swim, and hermit crabs still frolic. The best of the action is here.

DOGWINKLE SNAILS

Dogwinkle snails are predatory, carnivorous, and nocturnal, feeding in rocky intertidal areas on the ever-numerous barnacles and mussels. Because of their preference for the night, their activities are rarely seen by the casual observer, even though their several species have large populations. *Thaididae*

FRILLED DOGWINKLE
Nucella lamellosa (Gmelin)

The Frilled Dogwinkle snail thrives in practically all the rocky areas where there are crevices and other surfaces that provide shelter from direct wave action, providing a greater number and variety of shells for the beachcomber and diver than any other snail in the Northwest. It prefers hard, rocky surfaces but sometimes crawls across sandy or muddy surfaces where current or wave action is weak.

Frilled Dogwinkles may be dull orange, snow-white, brown, gray, or shades in between; solid in color or with white spiral bands. Bleaching, scrubbing, and coating with light mineral oil is often needed to remove a layer of algae and dirt and reveal the beauty of the shell.

Their shells are usually between 2 and 3 inches long but may reach 4 inches. Heavier shells of the open coast are thick and generally smooth, whereas the thinner shells in more protected areas often develop "frilling"—hence their name. Heavier shells also tend to develop several tooth-like bumps, around 3, in a row near the edge of the lip. Shell openings almost always form a point or nose containing a short canal.

To discover groups of resting FRILLED DOGWINKLES, like these at Devil's Elbow State Park, north of Florence, Oregon, one should look in crevices and beneath the protected side of large boulders. Within such groups are usually a few colorful ones like those shown on the following page, but their beauty is obscured by algae.

A group of FILE DOGWINKLES clustered in a crevice at Neptune State Park south of Yachats, Oregon. An extremely low tide in the spring often reveals groups of Dogwinkles with their egg cases as shown.

Color and patterns of the FRILLED DOGWINKLE, *Nucella lamellosa,* vary between locations and within locations.

Top: Left, intertidal rocks, Newport, Oregon. Center and right, shallow water, Nehalem Bay, Oregon.

Center: Left, intertidal rocks, Hood Canal, Washington. Right, intertidal, Sentinel Rock, San Juan Islands, Washington.

Bottom: Left, intertidal rocks, Tacoma, Washington. Center, 60-ft. depth near Johns Island, San Juan Islands, Washington. Right, piling, Gig Harbor, Washington.

The smaller dogwinkles also present a variety of colors and patterns, somewhat variable with locale.

Top: Left and center, FILE DOGWINKLE, *Nucella lima,* from intertidal rocks at Devil's Elbow State Park, Oregon. Right, CHANNELED DOGWINKLE, *Nucella canaliculata,* from same location.

Center: EMARGINATE DOGWINKLE, *Nucella emarginata,* from intertidal rocks at Lincoln City and Roads End, Oregon.

Bottom: EMARGINATE DOGWINKLE, *Nucella emarginata,* from intertidal rocks at the mouth of the Salmon River, Oregon (left) and Roads End (right). Note geographical differences.

EMARGINATE DOGWINKLE

Nucella emarginata (Deshayes)

A smaller relative of the Frilled Dogwinkle is the Emarginate Dogwinkle. It also has a variety of colors and shapes but is less widely distributed and lives primarily on rocky areas of the open coast, commonly among the mussels on the sides of large rocks. Shell colors may be orange, brown, yellow, gray, or ivory, often with about 6 single or double stripes spiraling around the shell. These shells usually require less bleaching and scrubbing to bring out their color than do many other shells.

Emarginate Dogwinkle shells from the Oregon beaches are usually short, globular, and smooth, while those from more northern areas—and one or two isolated, protected places in Oregon—are more elongated. The globular forms are not much more than an inch in length, but elongate ones may reach an inch and a half. Both forms may be either smooth or rough in texture, the smooth being more common. All Emarginate Dogwinkles have a characteristic large oval aperture with a small slit or canal.

FILE DOGWINKLE

Nucella lima (Gmelin)

File Dogwinkles occasionally appear in rocky areas of the open coast, though not as often as either the Frilled or Emarginate Dogwinkles. A good place to look for them is in large crevices and among the mussels, particularly on the sides of large rocks away from the heavy surf. The shell color of the File Dogwinkle varies much less than among other dogwinkles. It is usually white or beige, mottled with brown. The mottling tends to occur in bands.

In shape, the File Dogwinkle resembles a large Emarginate Dogwinkle, being globular with a similar aperture. Adult shells are commonly an inch to 1-1/2 inches in length, sometimes reaching 2 inches. They are characterized by about 18 smooth, rounded cords which spiral around the shell. Presumably these ridges gave rise to the common name of File Dogwinkle.

CHANNELED DOGWINKLE

Nucella canaliculata (Duclos)

Less common than the other dogwinkles is the Channeled Dogwinkle, which seems to prefer slightly deeper water. Thus it is more apt to be collected by skin divers than by beachcombers. Shell color resembles the File Dogwinkle—white with mottled brown in a somewhat banded appearance. Inside, though, it is often yellow. Texture also resembles the File Dogwinkle, but the Channeled Dogwinkle can be distinguished by its more elongate shape and the channels or deep grooves at the sutures (spiral seams between the whorls). It is also normally smaller.

The aperture of the Channeled Dogwinkle is oval but proportionately smaller and more elongated than that of the Emarginate and File Dogwinkles. The fold at the nose is also less pronounced.

DOVE SHELLS

The dove shells are a group of small, fairly common snails that reportedly feed on soft algae. They are not regularly seen because of their tendency to live below the intertidal zone and their habit of seeking shelter under rocks, sponge growth, or other material. *Columbellidae*

WRINKLED AMPHISSA

Amphissa Columbiana Dall

The Wrinkled Amphissa, most common member of the dove family in the Pacific Northwest, dwells in rocky areas and occasionally the intertidal zone. In beachcombing, it may be found under rocks at a particularly low tide and occasionally in the root-like clumps of holdfasts of kelpweed cast upon the beach during storms. A diver may find it under rocks and in crevices, usually where seaweed is abundant.

The Amphissa shell is small—rarely over one inch long—and slender, with the spire tapering to a sharp point. The latter half of the body whorl, and the smaller whorls, have fine lengthwise wrinkles; hence the name of Wrinkled Amphissa. Colors vary but tend to the orange browns.

Like many other snails of our coast, the WRINKLED AMPHISSA, *Amphissa columbiana,* develops thicker shells where exposed to wave action.

Top: Left, thick shell from Haystack Rock, Pacific City, Oregon. Right, intertidal rocks, Cape Arago, Oregon.

Center: Both specimens were found in the wreckage of the schooner *America,* San Juan Island, Washington.

Bottom: Both shells are from intertidal rocks, San Juan Island, Washington.

Coloration of the LEAFY HORNMOUTH, *Ceratostoma foliata,* seems to follow a geographical pattern, with darker shells being found farther north.

Top: Left, solid-colored shell collected subtidally at Henry Island, Washington. Right, a banded specimen collected intertidally at Cape Arago, Oregon.

Bottom: Left, pure white shell from Lone Ranch State Park near Brookings, Oregon. Right, banded specimen from San Juan Island, Washington.

MUREX SNAILS

The Murexes are carnivorous snails that prey on other mollusks, principally bivalves. Some are strong enough to seize a victim and pry it open; others drill holes in the victim's shell to eat the soft flesh beneath. An attractive characteristic of the Murex shells is a flaring out at the aperture into ridges or rows of spines at the end of each growth period. These ridges are called varices. *Muricidae*

LEAFY HORNMOUTH

Ceratostoma foliata (Gmelin)

Largest and most impressive of the Murexes found in the Pacific Northwest is the Leafy Hornmouth. Like most Murexes it has a fairly enlongated shell, sometimes reaching four inches in length. Also, similar to shells of other members of this family, each growth period ends in a flaring varix. These varices are broad and leaf-like, with a single spine at the base, next to the aperture—obviously providing the origin of the common name. Older varices remain as new growth periods extend the shell from the aperture, resulting in a shell with about three varices per turn of the whorl. Shell size and coloration appear to vary with location, with smaller white specimens typical in southern Oregon, brown-banded shells common along the central Oregon Coast, and solid-brown, larger shells occurring in northern Washington.

Since the Leafy Hornmouth is a driller of scallops, clams, and other bivalves, it appears where such a food supply exists. However, it avoids soft sand or mud, so one is most apt to find the Hornmouth intertidally in rock crevices, or subtidally to a depth of about 30 feet.

JAPANESE OYSTER DRILL

Ocenebra japonica (Dunker)

Tiny Japanese Oyster Drills were brought into the Pacific Northwest with seed oysters but have grown and become established in parts of Puget Sound, and in Netarts Bay, Oregon. Better adapted to soft substrate, they occur on tideflats where they aggressively feed on oysters, or in their absence, on clams.

Oyster Drill shells are elongated with 5 to 7 bluntly flaring varices. In larger shells the varix curls slightly at the end by the suture. Colors are shades of white and light brown with no distinct pattern, though there may be some appearance of banding. The varices are generally lighter in color.

The empty shells shown above have been drilled by predatory snails that suctioned out the meat of their victims. The scallop, lower right, was found being drilled by a Leafy Hornmouth; the others bear large holes typically left by Moon Snails.

LURID DWARF TRITON
Ocenebra lurida (Middendorff)

This small member of family Muricidae occasionally inhabits the under side of intertidal rocks but more commonly is found in deeper water. There, beneath the pounding surf, they live in greater numbers, at times hidden by areas of sponge growth and seaweed.

Lurid Dwarf Triton shells seldom exceed one inch in length, are rich brown in color, and textured with about 20 distinct spiral cords on each whorl. The aperture is relatively small, with a lip that often flares out from a row of little bumps resembling teeth. Varices are absent.

Ocenebra sclera (Dall)

Although this Dwarf Triton is larger, at times approaching 2 inches in length, it is uncommon enough not to have a common name. It is most apt to be found in subtidal rocky areas, principally along the islands of the Washington and British Columbia coasts. In addition to size, this Dwarf Triton is distinguished by roughly 8 to 10 rounded varices per whorl.

Top: JAPANESE OYSTER DRILL, *Ocenebra japonica,* collected from the tide flats at Netarts Bay, Oregon.
Center: LURID DWARF TRITON, *Ocenebra lurida,* specimens were found subtidally beneath sponge growth on the south side of Haystack Rock, Pacific City, Oregon.
Bottom: *Ocenebra sclera* were collected at a 20-foot depth near San Juan Island, Washington.

Oregon Triton, *Fusitriton oregonensis,* taken from a depth of around 15 feet at San Juan Island, Washington.

TRITONS

The tritons are a stately family of snails whose shells have attracted the attention of man for many centuries. Shells of the larger tropical tritons have been used as musical instruments in religious ceremonies. Tritons are said to feed on echinoderms such as starfish, and the author has observed the Northwest member of the triton family feeding on sea urchins. *Cymatiidae*

OREGON TRITON

Fusitriton oregonensis (Redfield)

The Oregon Triton is the Pacific Northwest representative of the triton family. Although it is not as large as many of the southern members of the family, it is still one of the Northwest's larger snails—sometimes over 5 inches long. Oregon Tritons occur in shallow water in rocky areas of northern Washington and in British Columbia, occasionally intertidally. Off the Oregon Coast they may be picked up in deep water by fishing nets, but are rarely found in shallow water.

The shells are fairly elongated with rounded whorls and distinct sutures. Lengthwise ridges, crossed by spiral cords, give the shell a woven appearance. Although the shell color is white, living specimens are covered with a brown, hairy layer called periostracum.

WHELKS

A wide range of snails with several shape characteristics is lumped into the broad common grouping of whelks. They include the large Knobbed and Channeled Whelks of the Atlantic Coast, many small whelks, and the neptunes. All are reportedly predatory. *Buccinidae*

DIRE WHELK

Searlesia dira (Reeve)

The Northwest's only shallow-water member of the whelk family, the Dire Whelk, is small, seldom exceeding an inch and a half in length. It occurs intertidally in rocky areas all along the open coast, and occasionally in sheltered waters. Although tending to live a little deeper than the dogwinkles, its habits are much the same—resting in the daytime and searching for food at night. The chocolate-brown shell of the Dire Whelk is elongated with a sharp spire and a pointed nose. Earlier whorls have a number of fold-like ridges lengthwise to the shell, but these are lacking in the adult body whorl. Spiral cords are apparent inside as well as outside.

HAIRY SNAILS

Hairy Snails, as cold-water dwellers, include the northern part of the Pacific Northwest within their range. Although they wear a similar-appearing brown furry periostracum, they are not closely related to the Oregon Triton (also called the Hairy Triton). While the Oregon Triton is a predator, the Hairy Snail feeds upon marine algae. *Trichotropidae*

CHECKERED HAIRY SNAIL

Trichotropis cancellata Hinds

Checkered Hairy Snails are sometimes seen in rocky areas, mostly along the shorelines of northern Washington and British Columbia, but are most apt to appear on the under side of subtidal rocks. Like the Oregon Triton, the Checkered Hairy Snail has a brown tufted covering over a thin, white shell. However, the Checkered Hairy Snail is small, generally not over an inch in length, and differs from young Oregon Tritons with its rounded aperture and lack of folded ridges on the whorls.

Top and center: Specimens of DIRE WHELK, *Searlesia dira,* collected intertidally at Whale Cove, Oregon.

Bottom: CHECKERED HAIRY SNAIL, *Trichotropis cancellata,* found among shallow-water rocks at Smallpox Bay, San Juan Island, Washington.

Top: The BLACK TURBAN, *Tegula funebralis,* left, was collected from intertidal rocks in Whale Cove, Oregon, while the DUSKY TURBAN, *Tegula pulligo,* right, was plucked from Kelp fronds in the same cove.

Bottom: BLUE TOP, *Calliostoma ligatum,* shells from shallow water at San Juan Island, Washington.

TOP SHELLS

Top Shells include a number of different genera of snails whose common characteristics are that they are herbivorous, scraping up bits of marine algae, and have a general shape of a flattened or rounded aperture, sloping sides, and pointed spire—not unlike a child's spinning top. A pearly interior layer has caused many, like the tropical Commercial Trochus, to be used in the manufacture of buttons. In life, the snail travels with the spire upward. *Trochidae*

BLUE TOP
Calliostoma ligatum (Gould)

The Blue Top is a small, colorful inhabitant of rocky areas, often found on the low-growing Ribbon Kelp (Laminaria). It is named for the bright-blue pearly interior layer of the shell, which shows through when the shell becomes slightly worn from water action. Younger ones, or those living in sheltered places, are light brown. The blue color, however, is what is most attractive to collectors.

The shell aperture, and the sides of the whorls, are rounded. Twelve to 15 raised, smooth ridges spiral with the whorls. There is no nose, canal, or umbilicus. Height is normally less than one inch. (Height, often used to designate shell length, best applies to measurement of Top Shells.)

BLACK TURBAN
Tegula funebralis (A. Adams)

The Black Turban is a common snail of the intertidal rocks of the open coast. During low tides, groups of them are often seen clinging in niches of rocky outcroppings. In these outcroppings, exposed to the surf, the black outer layer of the older whorls is normally worn away, showing a dull but slightly pearly inner layer.

Although it is a top shell, the sides of the whorls and the spire of the Black Turban are smooth and rounded, with only a thin line at the sutures. The aperture is also rounded, with two small tooth-like bumps close to the indentation of the shallow umbilicus. Adult shells are usually a little over an inch in height and slightly wider than high.

DUSKY TURBAN

Tegula pulligo (Gmelin)

The Dusky Turban is not a true rock dweller. The author has more often found it clinging to the fronds of giant Bull Kelp as they swayed gently in the swells. Although larger than the Black Turban, it climbs easily but will often drop to the ocean floor when disturbed—later climbing back to its former browsing area.

The shell of the Dusky Turban is flat sided and conical. As the whorls coil with added growth, they do not completely meet in the center, leaving an open hole, called an umbilicus. The exterior of the Dusky Turban is dull orange or beige, the interior pearly white.

WENTLETRAPS

This group encompasses a large number of species of small, usually slender, and predominantly white shells with numerous delicate lengthwise ribs crossing the whorls. The name stems from the Dutch word for spiral staircase. Such shells are often highly valued by collectors, particularly the tropical Precious Wentletrap, which was once so expensive that replicas were made of rice paste. *Epitoniidae*

MONEY WENTLETRAP

Epitonium indianorum (Carpenter)

These delicate little shells (generally less than one inch in length) are uncommon and not regularly found in any area. However, they feed upon anemones and occasionally appear where anemones are abundant, at times sharing an intertidal area with the more common Opal-Shell.

Characteristics of the Money Wentletrap are the slender, white appearance and usually 13 or 14 sharp, encircling, collar-like ridges around each whorl. The round aperture, except during periods of growth, flares out in the shape of one of these collars. Sutures between whorls are distinct.

Top: The NORTHERN OPAL-SHELL, *Opalia chacei,* was found in a rocky tidepool near Otter Rock, Oregon.

Center: MONEY WENTLETRAP, *Epitonium indianorum,* from a crevice in intertidal rocks at Devil's Elbow State Park near Florence, Oregon.

Bottom: This THREADED BITTIUM, *Bittium eschrichtii,* was taken from among the thousands that live under intertidal rocks of the south jetty to Yaquina Bay, Oregon.

NORTHERN OPAL-SHELL

Opalia chacei Strong

The Northern Opal-Shell is larger than the Money Wentletrap, often reaching over one inch in length. It is also more common. A similar species, Wroblewski's Wentletrap, is found in British Columbia and Alaska. Both Opal-Shells are occasionally discovered in rocky intertidal areas of the open coast. Collectors of these shells are occasionally surprised by stains on their fingers, for wentletraps like these exude a purple dye when disturbed.

Opal-Shells are slender and white but lack the delicacy of some wentletraps. The ribs are thicker and flattened, as are the whorls; the collar at the aperture is indistinct compared with the Money Wentletrap.

BITTIUMS

These snails are often overlooked because of their small size and life style, which usually keeps them under rocks and stones. They live in groups, feeding upon the residue that washes into crevices between rocks in the intertidal zone. Bittiums belong to the large scientific family of Horn Shells. *Cerithiidae*

THREADED BITTIUM

Bittium eschrichtii (Middendorf)

Although the Threaded Bittium is barely over a half inch in length, it is large for a bittium. In fact, another name for it is Giant Pacific Coast Bittium. Groups of them are readily discovered by turning over rocks in most coastal rocky points, and often in estuaries. (Any rocks overturned while collecting should be returned to their original position to prevent the death of the organisms adapted to life on the underside.)

The shell of the Threaded Bittium is slender with nearly straight sides and indistinct sutures. It is textured by spiral grooves or threads interrupting a smooth surface and colored a dull gray or brown. The aperture is oval with a curling that forms a slight nose.

Top: Two rows of CHECKERED PERIWINKLE, *Littorina scutulata,* those in the uppermost row from an abundant supply on intertidal rocks at Port Ludlow, Washington; those in the second row picked from Rockweed at Three Rocks, Oregon.

Bottom: The four specimens of SITKA PERIWINKLE, *Littorina sitkana,* were also found on Rockweed at Three Rocks.

PERIWINKLES

The tiny periwinkles have almost ceased to be marine snails. They live so far up in the intertidal zone that they go for long periods completely out of the water, leading some scientists to theorize that our land snails evolved from such an animal. *Littorinidae*

CHECKERED PERIWINKLE
Littorina scutulata Gould

The Checkered Periwinkle is normally less than a half inch in length. Its elongated and tapering shell ends with a sharply pointed spire at one end and a teardrop-shaped aperture at the other. The sides are straight with indistinct sutures.

Shell colors and patterns vary occasionally, according to particular geographic areas. Basically, the exterior color is dark brown to nearly black, often with white dots. At times the larger dots appear in fascinating patterns. These shells may also have spiral bands of dull orange with cross bars or spots of white. Interior colors vary, though dark purple predominates in the darker shells.

SITKA PERIWINKLE
Littorina sitkana Phillippi

The larger Sitka Periwinkle, whose shell is often over a half inch in length, is common in the northern part of the Pacific Northwest and appears along the Oregon Coast. It is usually found living on Rockweed or in nearby crevices. The shells of the Sitka Periwinkle are distinguished by their somewhat globular shape, obvious sutures, and distinct spire. The sides of the whorls are rounded and textured with spiral cords that may or may not be raised.

Colors and patterns vary greatly on these shells, although the basic seems to be some shade of brown. Lighter brown, yellow, or white bands often decorate the shells, forming either one broad or two narrow stripes around the shell. These colorful shells are attractive, and were they only larger, would certainly be favorites among collectors.

ABALONES

The abalone is a true snail, though its aperture is so large the shell is often mistaken for a bivalve. Unless it is too eroded or riddled by borers, the shell will reveal a flattened spire. The abalone leads a rather sedentary life, feeding upon marine algae that it scrapes from rocks with its radula (rasp-like tongue). In turn, it is fed upon by crabs, otters, and octopi. Occasionally a hoard of abalone shells may be gathered from an octopus den where they were discarded after the octopus dined. *Haliotidae*

PINTO ABALONE

Haliotis kamtschatkana Jonas

The Pinto is the more northerly dweller of the abalones. It is abundant from northern Washington to Alaska in locations where there is a flow of clear ocean water and plenty of kelp growing from rocky formations. Although occasionally found intertidally, the author has encountered the greatest numbers about 20 feet beneath the surface, and occasionally at 70 feet. The flesh of this species has fine eating quality.

The shell of the Pinto Abalone commonly measures about 5 inches at the widest point of the aperture, occasionally as much as 7. It is more oval in appearance than the southern Red Abalone and deeper (more concave) than either of the other abalones of the Pacific Northwest. Although thin and fragile, the typical shell has a number of wrinkle-like ridges radiating toward the edge of the shell. Some spiral cords are usually present, though not as numerous nor as distinct as on the shells of other Northwest species. This texture, plus the number of open ports (usually 4 or 5) is the distinguishing characteristic.

Exterior shell coloration is delightfully varied. Most common is red, either solid or with bright patterns of blue or white, but green, burnt orange, and other combinations are not unusual. Its bright, pearly interior was used for shell inlays in Northwest Indian art.

Interior view of PINTO ABALONE, *Haliotis kamtschatkana,* showing the bright, pearly portion used for shell inlays in Indian art.

Exterior view of the same PINTO ABALONE shell, taken from approximately 20-foot depth, Sentinel Rock, San Juan Islands, Washington.

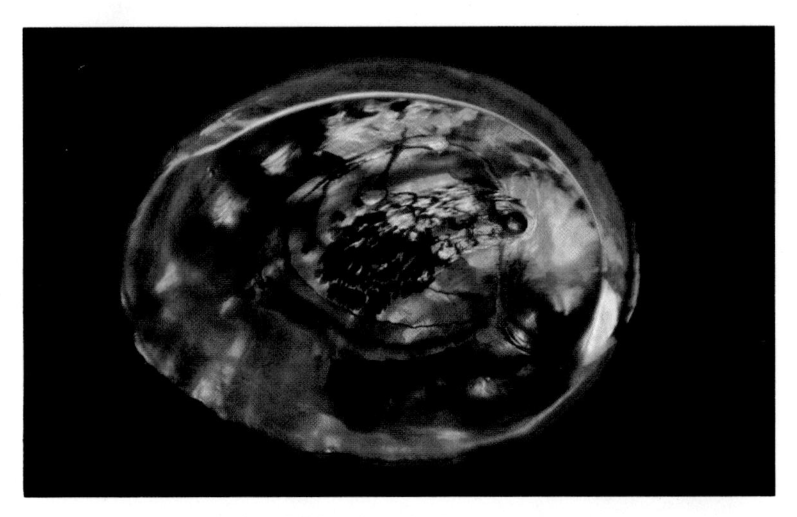

Interior view of a RED ABALONE, *Haliotis rufescens,* taken from shallow water rocky area near Brookings, Oregon.

Exterior view of a RED ABALONE, *Haliotis rufescens,* from the same area. This shell has a research tag in place, as it was attached to the living animal.

RED ABALONE

Haliotis rufescens Swainson

The Red Abalone is important to the Pacific Northwest because it is an edible shellfish and has the largest and most impressive shell. The natural range of this snail is only as far north as Coos Bay on the Oregon Coast. However, efforts have been made to introduce spawning stock at several locations along northern Oregon and Washington.

Shells of Red Abalone collected along the southern Oregon Coast often reach a width up to 11 inches. This may result from a less crowded environment—fewer people and sea otters to prey on the abalone areas, in contrast to the California Coast. Despite their size, Red Abalone are usually difficult to find, requiring the collector to reach deeply into tidepool crevices at extremely low tides, or to skin-dive along offshore reefs. When found, they are commonly covered with algae, barnacles, and other encrusting matter, making them appear like rounded, flat rocks.

The shell is oval, though it is the most nearly round of Northwest abalones. Texture is relatively smooth except for encircling lines of growth. Larger shells may also develop a number of radiating wrinkles. As the shell grows, the abalone seals its ports in the older part of the shell and forms new ones in the growing edge, usually maintaining (in the case of the Red Abalone) a row of four open ports for tentacles or feelers.

Exterior shell color of the mature Red Abalone is a brick red. During initial stages of growth it is often green, attributed to diet. This green area of the spire may continue as the abalone grows, depending on shell abrasion. The most attractive part of this shell is the iridescent, pearly interior. Blushes of olive green add to the beauty, occurring in greatest intensity in the area of the muscle scar, which is the irregular but roughly circular portion with lined patterns near the center of the shell.

FLAT ABALONE

Haliotis walallensis Stearns

The Flat Abalone is not as common as the Red but does occur in various places all along the Northwest Coast. Like the Red Abalone it lives in rocky areas, but is not generally found in the intertidal zone. Consequently, it is more often found by skin divers than by beachcombers.

The shell is smaller than that of the adult Red Abalone, generally not exceeding 5 inches at the widest point. It is usually more elongated and much flatter (less concave) than either of the other abalones. When undisturbed, the broad, thin mantle of the animal contributes to the "flat" appearance, extending well past the edge of the shell. Even the row of ports left open for the feelers or tentacles is not as raised and distinct. The number of these open ports varies, but is normally 6 to 8.

Exterior coloration of the Flat Abalone is a dark brick-red, often with white or blue-green markings. The surface is textured with indistinct growth lines and many evenly spaced, distinct cords radiating out from the spire. This cording is also apparent on the inside of the thinner shells.

The interior of the Flat Abalone shell is pearly with a silvery iridescence, like the other abalones. However, while the Red Abalone always has a noticeable muscle scar and the Pinto has muscle scars in the old and/or large shells, the Flat Abalone rarely shows a scar at the site of muscle attachment. Its shells, however, are more prone to infestation of piddocks and other boring bivalves. These pests bore through from the outer surface of the abalone shell, obliging the creature to build up a shelly layer of protection. This appears at times like large pearls, other times as mere bumps, and occasionally as blackened blisters on the interior surface. Only a small hole, about the size of a pencil lead, reveals the borer's presence in the exterior side of the shell.

Shells of FLAT ABALONE, *Haliotis walallensis,* taken from a shallow-water rocky area near Port Orford, Oregon. The interior view shows the typical absence of a muscle scar.

LIMPETS

While the abalone no longer resembles the spiral theme of a shelled snail, it still carries a vestige of a coiled spire. In the case of the adult limpet snails, the coil is lost, being present only in some of the minute, very young shells. By adulthood the Keyhole Limpet has evolved a single hole at the apex of the shell (in addition to the aperture), for a different purpose, but somewhat analogous to the holes in the abalone shells. Other limpets have only rather flat, uncoiled shells. All limpet snails are browsers of marine algae; most are intertidal. *Acmaeidae and Fissurellidae*

WHITECAP LIMPET
Acmaea mitra (Rathke)

The Whitecap Limpet shell is commonly washed up on sandy beaches along the open coast, appearing white and conical like a little duncecap. It is, however, the shell of a snail that lives on rocks, the empty shells being brought to the beach by wave action. Since the whitecap is deeper dwelling than most Northwest limpet snails, it commonly resides on offshore reefs. The living Whitecap Limpets may be seen by skin divers, and occasionally may be encountered in rocky tidepools at extremely low tides.

Like other limpet snails, Whitecap Limpets have shells that blanket their backs rather than spiraling around their body. This permits the soft parts of their anatomy to be together instead of strung out in a spiral form. Like other forms of shells, though, the shell grows with the animal by the addition of material to its edge.

The shell of the Whitecap Limpet is white, conical, and smooth, but living specimens tend to collect encrustations that give them a rough appearance and a variety of hues. (This is usually abraded on shells washed up on sandy beaches.) On the inside of the shell is a tiny horseshoe-shaped area which is the muscle scar where the snail was attached to its shell. The broad aperture at the bottom is nearly round and may be up to an inch and a half wide. These shells are usually wider than they are high.

MASK LIMPET

Notoacmaea persona (Rathke)

The Mask Limpet is a common intertidal inhabitant of the Northwest Coast. Since it dwells high in the tide zone it can be observed or collected on any low tide. There it lives and browses in a restricted area of the rocky surface, at times wearing a "home spot" into the site where it rests.

The shell of the Mask Limpet is wide and oval, quite limited in height. The apex or point in the center of the shell is flattened, and the sides slope in a rounded manner. Texture is nearly smooth, with indistinct circles of growth and slight ridges radiating out from the apex. Exterior color is dark brown or black, liberally specked or splotched with light gray dots which occasionally seem to be in a pattern like a witchdoctor's mask. Inside coloration is white with a narrow black band around the edge, and at times a brown blotch in the center. Larger shells in some areas may reach 2 inches in length, but most are much smaller.

UNSTABLE LIMPET

Collisella instabilis (Gould)

The Unstable Limpet has made an interesting adaptation to its environment. It is not a true rock dweller, but spends its life on the stipes (stalks) of the Kelp Weed that anchor to the shallow-water rocky areas. The shape and color of the Unstable Limpet have evolved so that it looks like a lump on the stem of the kelp. Skin divers can observe Unstable Limpets by examining the normally smooth stems under water. When the snail dies or is eaten, the shell falls to the ocean floor, where it is often incorporated into the development of the root-like anchor or holdfast of the kelp. Subsequently, it may be carried to the beach with the kelp that is washed ashore. That is where beachcombers find it.

The shell of the Unstable Limpet is basically oval, but the edges are distorted in a wrap-around manner to conform to the curvature of the stipes of the kelp. The shell is olive brown with a white interior. Length may reach slightly over an inch.

FINGERED LIMPET

Collisella digitalis (Rathke)

The Fingered Limpet is common on intertidal rocks all along the Northwest Coast. Like the Mask Limpets and the periwinkles, colonies of them live high in the tide zone where they can be seen during any low tide. They can usually be found in the shaded areas of open rock.

Fingered Limpet shells are small, usually not over an inch in length. Their shape is conical, with an oval aperture, not entirely symmetrical. With the apex forward of center, the front slopes steeply, while the reverse is long and rounded, like the crown of a fireman's hat. Radiating back from the apex are a number of ridges or fingers, as if a tiny hand were placed on the tip of the cone with the fingers extending down the sides to the edge. The exterior shell is gray with brown zigzag markings; the interior white with a brown blotch in the center and darker brown band along the edge.

ROUGH KEYHOLE LIMPET

Diodora aspera (Rathke)

As with several other rock dwellers, the shells of the Keyhole Limpet are frequently washed from offshore reefs to beachcomber areas. Normally residing deeper than most limpets, the living animals are best observed by skin divers. The shells—so brightly striped when their sanded remains are found on the beach—are difficult to see in the living animal because their rough surface is camouflaged with a layer of algae. Near the apex of the shell is an oval port or "keyhole" somewhat like one of the ports of an abalone. This hole, which moves in from the edge of the shell during infancy, serves as a vent for excretion.

Rough Keyholes are large for limpets, reaching 3 inches in length and attractively marked with about a dozen black stripes crossing the white shell from apex to border. Their size and markings make them collectors' items.

Top: Shells of the WHITECAP LIMPET, *Acmaea mitra,* collected subtidally at Whale Cove, Oregon. (The shell on the left is encrusted with coraline algae.)

Center: MASK LIMPET, *Notoacmaea persona,* left, and FINGERED LIMPET, *Collisella digitalis,* right, from intertidal rocks at Cascade Head, Oregon.

Bottom: The UNSTABLE LIMPET, *Collisella instabilis,* left, was removed from the stem of a Kelp Weed in Whale Cove, Oregon; the ROUGH KEYHOLE LIMPET, *diodora aspera,* right, washed ashore at Beverly Beach, Oregon.

Sand Dwellers

Moon Snail near Quilcene, Washington, digging a clam.

While a variety of sea animals is readily apparent in rocky tidepools, one must seek out the more elusive dwellers of the sandy and silty areas. Along open beaches the pounding surf has obliged the residents to dig beneath the sand for protection. In muddy estuaries and sheltered waterways the wave action is absent, but similar escape patterns cause residents to go beneath the soft silt to elude predators and avoid drying and temperature change when the tide is out. Often it is only their tracks or a slight mound that indicates their presence.

In this bleak-appearing but rich community, large populations of filter-feeding bivalves, digging crustaceans, and marine worms thrive. With them lives a variety of snails. Some, like the abundant olives and the less-common nassas, exist as scavengers; but the moon snails, which are impressive both in life style and size, are predators. Yet all have adapted a wide and powerful foot to carry them across or through the soft sand or mud. The moon snail has developed this locomotion to the ultimate, gliding rapidly over the silt and digging several inches beneath the surface to capture clams.

MOON SNAILS

Moon snails, whose shells are round like a full moon, are some of the largest sand-dwelling snails of the Pacific Northwest. They are also giants when compared with fellow moon snails of other parts of the world. Because they are active in the daytime, crawling over sand and silt, digging clams, and, at times, building sand collars in which to deposit their eggs, they are interesting for the skin diver to observe. *Naticidae*

LEWIS'S MOON SNAIL
Polinices lewisii (Gould)

Lewis's Moon Snail is a large snail of the tide flats and shallow water, at times reaching a size of 5 inches across the largest dimension of the shell. Under water, the large, broad foot extends well past the edges of the shell and the mantle extends up far enough to sometimes completely cover the shell, making it appear like a huge, yellowish slug. This big foot enables the snail to move rapidly over the soft ocean floor and to dig rapidly into the mud or sand. It also enables the snail to hold victims for rasping a hole into the shell to consume the occupant. The author has watched the Lewis's Moon Snail as it dug out of sight in the silt to get a clam, as it glided over the soft bottom on a foot so large the mantle covered the shell, and as it lay on its back with the foot raised, manipulating the "white-sidewall-looking" sand collar.

When the tide is out, the heavy mantle is cumbersome and falls away from the shell. At this time the moon snail normally buries itself in the soft mud—but can occasionally be spotted by the mound where it entered.

Like others of its family, the shell of the Lewis's Moon Snail is round like a full moon. Sutures are obvious, but the indentation is not so great as to detract from the round appearance. The aperture, with its rounded outer edge and straightened inner edge, is semicircular. The whorls do not completely meet, so there is a deep umbilicus, bordered by a shelly deposit, or callus, at the edge of the aperture. Since the siphonal fold, which fits into the canal in the shells of most predatory snails, is carried outside in the huge mantle, the moon snail shell lacks the canal that characterizes most predatory snail shells.

ARCTIC NATICA

Natica clausa Broderip & Sowerby

Occasionally, particularly from Puget Sound northward, one finds a smaller moon snail, the Arctic Natica. It is more elusive than the Lewis's Moon Snail, frequently remaining buried during daylight hours. It is also reportedly a deeper dweller, but its shells do get carried to beaches occasionally by strong currents and by hermit crabs. Numerous juvenile shells have been found in bottles that were occupied by octopi and later picked up by skin divers. Most living specimens that are found, however, are dug out during clam digging.

Like Lewis's Moon Snail, the shell of the Arctic Natica is round. It can be distinguished from the former by the absence of an umbilicus—this is filled by a large white callus. It can also be distinguished, in the case of living specimens, by the type of operculum. Despite their voluminous bodies, moon snails can, by expelling large amounts of water, withdraw completely into their shells. They then seal the aperture with an operculum or trap door. Most of our moon snails, like the Lewis's, have operculums of an amber, translucent material. The Arctic Natica, instead, has an operculum made of white, opaque shell material. Whorls are also more rounded and regular and of a smoother texture.

Shell color of adults is a light beige, inside as well as out, except for the callus and adjoining white area. Small juvenile specimens, about the size of peas, are chestnut brown in color with lighter, irregular stripes. By the time they have reached the size of a marble, this colorful striping unfortunately is lost, and a uniform beige, darker than the adults, replaces it. The size of the adult Arctic Natica shell is small compared to the Lewis's Moon Snail, the Natica normally being no larger than one inch.

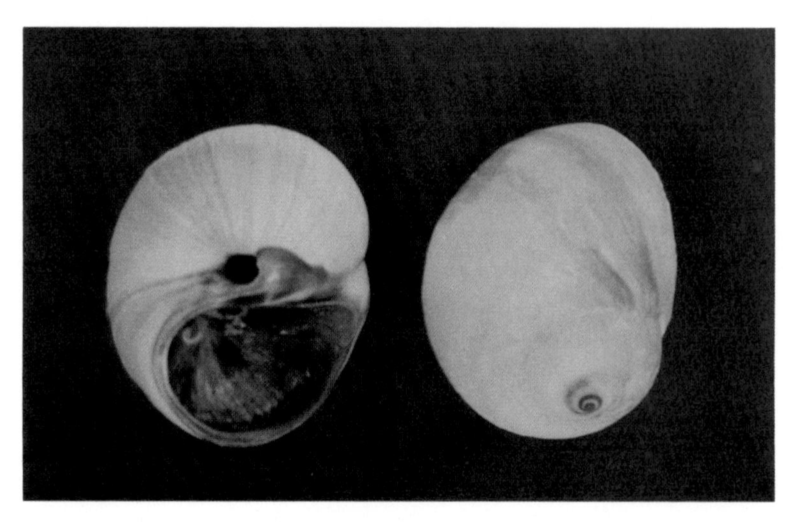

Shells of the LEWIS'S MOON SNAIL, *Polinices lewisii,* gathered from a depth of 20 feet near Seal Rock (Hood Canal), Washington. Note the operculum or trapdoor closing the aperture.

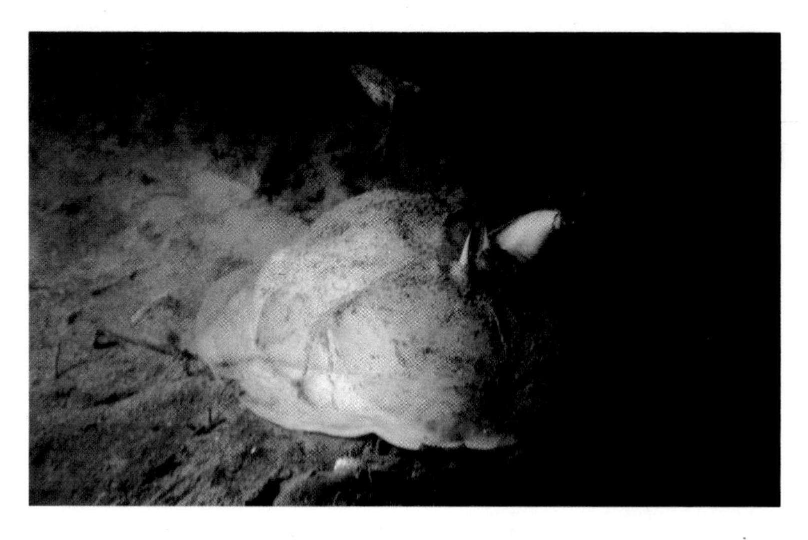

The foot is so wide on the LEWIS'S MOON SNAIL, and the mantle so large, that the shell is nearly hidden on this one photographed under water near Quilcene, Washington.

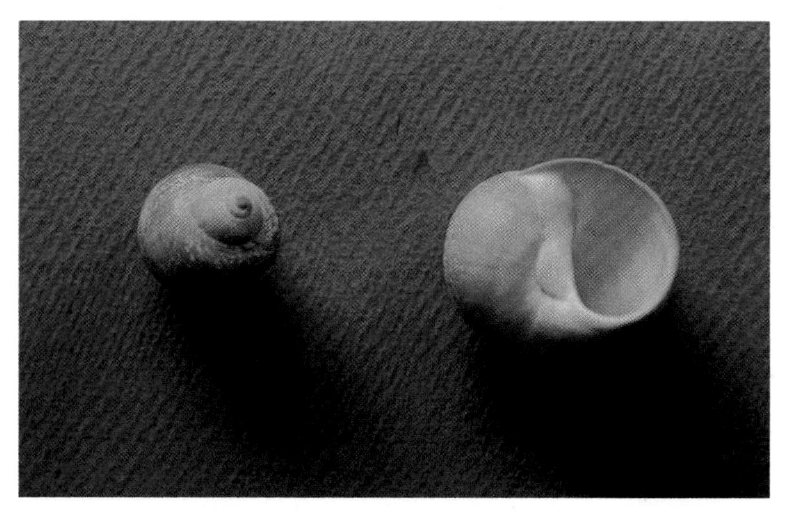

ARCTIC NATICA, *Natica clausa,* shells found subtidally at Smallpox Bay, San Juan Island, Washington.

These moon snails are deeper dwellers and were obtained from shrimp nets near Newport, Oregon.
Top: DRAKE'S MOON SNAIL, *Polinices draconis.*
Bottom: NORTHERN MOON SNAIL, *Polinices pallida.*

DRAKE'S MOON SNAIL

Polinices draconis (Dall)

Drake's Moon Snails are occasionally picked up by skin divers in offshore areas. Because they live in deeper water than most of our moon snails, however, their shells are rarely washed up on the beach. They are important predatory snails in the flats of the continental shelf, as the Lewis's Moon Snails are in the estuaries and sheltered waterways.

The shells and operculums of the Drake's Moon Snails are much like those of the Lewis's. Both shells are rounded and colored a creamy beige; both operculums are of a plastic-like horny material. The Drake's Moon Snail is smaller, usually no larger than 3 inches across, and somewhat flatter. The umbilicus is its most distinctive characteristic. The innermost side of the shell, which faces the umbilicus, is thinner. This wall is mismatched and inset from the rest of the curving shell. As a result, there remains a narrow shelf of the older whorls showing in the umbilicus, giving an effect like looking down into a narrowing spiral staircase.

NORTHERN MOON SNAIL

Polinices pallida (Broderip & Sowerby)

Like the Drake's Moon Snail, the Northern Moon Snail lives offshore on the continental shelf. It is sometimes discovered by skin divers, but is usually overlooked because of its small size. Most collectors obtain their specimens of the Northern Moon Snail from shrimp fishermen.

The shells of this species look like other moon snails, but are much smaller. Adult shells are less than one inch across. In shape, they are more perfectly round, both in the outline of the whorls and across the aperture. The absence of a callus and the near-absence of an umbilicus add to the smooth, round appearance of these little shells. A thin periostracum coats the shells on the living snails and contains a yellowish gray pigmentation. When the snail dies, the periostracum peels off, revealing a white shell beneath. This whiteness accounts for the other name for this creature, the Pale Moon Snail.

OLIVE SNAILS

Olives are the most abundant scavenging snails of the beaches and estuaries. Although thousands of them inhabit single tide flats, their life style of digging beneath the sand in search of food leaves them almost undetected. Their shells are characterized by elongated oval whorls, nearly absent spires, and narrow slits of apertures, ending with a notch canal in a wrinkled nose. They are at times colorful, always glossy and smooth in appearance. *Olividae*

PURPLE OLIVE

Olivella biplicata (Sowerby)

The Purple Olive is the most common snail of the tide flats, sandy coves, and open beaches. In quiet areas they can often be found by the trails they leave as they plow along under the sand or mud; in the breakers they may be seen sailing about with the surging sand particles, almost skater-like as they glide about with the waves.

Olives have a wide digging foot and a mantle that covers the shell. It is this mantle that keeps the glossy finish on the shells. The Purple Olive has an oblong, blunt-ended shell with very little spire. A few fine wrinkles are found in the nose which appears folded. These are less distinct, however, than those of larger tropical olives. Colors are normally shades of blue-gray, though ivory or brown shells are found in some places. A narrow purple band outlines the nose, which is typically lighter. Length is less than one inch.

BEATIC OLIVE

Olivella baetica Carpenter

The Beatic Olive shares the same sandy coves and tide flats as the Purple Olive, but is neither as common nor as large. It is reported that while this is true in Oregon and Washington, the Beatic is the only olive found in significant numbers as far north as Alaska.

The shell of the Beatic Olive is oblong, but has a longer, more tapering spire. Sutures are more distinct and the shell is usually more colorfully marked (although plain specimens are also found). Typically, they are cream with zigzag patterns of brown. Length is normally about a half inch.

THE NASSAS

Also called dog whelks and mud snails, the nassas are a family of small, stubby snails that prowl the sand and mud of sheltered coves, estuaries, and inland waterways in search of food. They are classified as carnivores, but the author has observed them as scavengers, similar in some habits to the olives. *Nassariidae*

GIANT WESTERN NASSA
Nassarius fossatus (Gould)

Large populations of this snail inhabit the sandy coves and sheltered beaches along the Northwest Coast. Like the olives, the Giant Western Nassas dig beneath the sand, but living in a wave-washed environment, their trails are seldom seen. Because their shells are bulkier and more easily caught up by strong surf, they must avoid open areas that support colonies of competing olives. At Devil's Elbow State Park, near Florence, Oregon, the author has seen dozens of these snails cast up on the beach by southwest storms.

Although larger than most nassas, the shell of the Giant Western Nassa reaches a length of only about an inch and a half. It tapers rapidly from a pointed spire, has a body whorl somewhat flattened on the forward side, and ends in a stubby, up-turned nose. This upturned nose, with a groove around its base, ends in an upturned notch canal, characteristic of all nassas. Another characteristic is the parietal shield, a callus-like extension on the body whorl from the aperture.

The sculpture and colors of the Giant Western Nassa make this one of our most interesting intertidal shells. It is wound with about 20 spiral cords and crossed with numerous fine ridges that give the shell a basket-like appearance. (It is sometimes called the Basket Shell.) The ridges, however, do not continue to the forward, flattened portion of the body whorl, so the basket weaving is incomplete. The basic color is tan, in some cases (particularly beach-worn shells) quite light, in other variations dark. The aperture and parietal shield are commonly bright orange, the interior striped with the effect of the spiral cords.

Top: GIANT WESTERN NASSA, *Nassarius fossatus,* collected intertidally at Devil's Elbow State Park, Oregon.

Second Row: WESTERN FAT NASSA, *Nassarius perpinguis,* left, and LEAN NASSA, *Nassarius mendicus,* found in Yaquina Bay, Oregon.

Third Row: Two specimens of PURPLE OLIVE, *Olivella biplicata,* collected intertidally in Yaquina Bay.

Lower: Specimens of BEATIC OLIVE, *Olivella baetica,* same location.

WESTERN FAT NASSA
Nassarius perpinguis (Hinds)

One may encounter the Western Fat Nassa along the Oregon and southern Washington coasts. It also lives beneath the surface of the sand, though the author has found it prowling about on top of the mud in the channels far up the Yaquina estuary. Most commonly, however, the shells are found being used by hermit crabs in sandy tidepools and tide flats.

The Western Fat Nassa is smaller than the Giant Nassa, but is slightly wider in proportion to over-all length. It seldom is over an inch long. Whorls are rounder, lacking the flattened effect, and are textured with more distinct ridges, giving the shell a banded appearance. The aperture is wide, but the lip is not as thick and protruding and the parietal shield not as noticeable. Colors are more drab, with white rather than orange around the aperture.

LEAN NASSA
Nassarius mendicus (Gould)

The third nassa commonly found in the Pacific Northwest is the Lean Nassa. It lives in a wide variety of places, but is most common in tide flats and sheltered waterways, sharing the range of the Fat Nassa and also residing in northern Washington and British Columbia. Under water it seems to spend more of its time above the surface of the mud, so is more often seen than its two relatives.

The shell of the Lean Nassa is occasionally an inch in length, more commonly about three quarters. Whorls are rounded and separated by distinct sutures. It is a much more slender shell, not broadening at the body whorl like its two Northwest relatives. The texture includes about a dozen spiral cords, but the numerous fine ridges are replaced with about 15 wrinkle-like ridges. (A separate sub-species, called *Nassarius mendicus cooperi*, and reportedly only in Washington, Oregon, and southward, has fewer ridges and is smaller.) Shell colors are brown and ivory, occasionally white, with one or more bands of brown spiraling around the shell.

Deep-water Dwellers

Observations while walking along the beach, particularly at low tide, can provide information about olives, nassas, and other sand dwellers. By exploring the tidepools, and by skin diving we can extend our observations to the dogwinkles, murexes, limpets, and their neighbors. Even with scuba equipment, however, we can only touch the fringe of another area in which many of our most fascinating shelled animals live—the deep-water zone.

In this zone, several hundred feet deep, large snails live. Freed of the problems of wave action, they are able to develop large and fragile shells. Within the dark depths adjoining the Pacific Northwest there are a number of such snail species, snails that produce some of our most noteworthy though drably colored shells.

These shells, however, are more difficult to obtain. Residing below the normal wave action, they are rarely washed to the beach intact. Occasionally a shell may be carried to shallower water by a hermit crab, and one species reportedly goes to milder depths to lay its eggs. On these occasions a skin diver may be surprised with a treasured find. Usually, though, shells of the deep dwellers must be dredged for or obtained from fishermen who sometimes find shells, such as those of neptunes, entangled in nets used to gather deep-water fishes for market. A few collectors even search through stomach contents of freshly caught deepwater fish for these shells.

NEPTUNES

The neptunes belong to the extensive family of whelks (Buccinidae), a group of large carnivorous snails found mostly in northern seas, but whose family also includes the Channeled and Lightning Whelks of our warmer Atlantic states. The neptunes, which once were afforded their own family, Neptuneidae, are all found in colder, deeper waters. Few of the Northwest species are found even within the range of skin divers. *Buccinidae*

PRIBILOF NEPTUNE

Neptunea pribiloffensis (Dall)

The impressive shells of the Pribilof Neptune are occasionally encountered by skin divers. Such discoveries are normally of dead shells occupied by hermit crabs, for the living snails inhabit the depths. Since shrimp and bottom-fish nets that operate in water a couple hundred feet deep often entangle and bring up these snails, most collectors seek out fishermen as a source of the shells.

Pribilof Neptunes become quite large for cold-water shells, at times 5 inches in length and 3 in width. Their shape is fairly elongated, like most predatory snails, with a distinct spire and a canal in the nose. Whorls are full and rounded with distinct sutures where they join. This is a bulky snail with a large body mass, so needs a large aperture and voluminous body whorl to withdraw into. When inside, it closes the opening with a large teardrop-shaped operculum of horny (plastic-like) material.

The shells are smooth, although earlier whorls have a few fairly distinct spiral cords. Less noticeable are the lengthwise growth lines. Since the shells are thin and relatively fragile, occasional irregular growth-like lines may be present. These are actually places where the edge of the shell was damaged and repaired by subsequent growth. While the size and graceful shape of the Pribilof Neptune shell can be considered impressive, the texturing and the coloration could hardly be regarded as anything but plain. The soft, yellowed white or faint beige color is not particularly interesting.

An average-sized specimen of PRIBILOF NEPTUNE, *Neptunea pribiloffensis,* this 3-1/2-inch shell was scooped up in a fishing net of the otter-trawl type offshore from Tillamook Head, Oregon. Water depth at that location is about 200 feet.

TABLED NEPTUNE

Neptunea tabulata (Baird)

The Tabled Neptune is another member of the Whelk family. Like the Pribilof Neptune, it lives in the sandy and silty areas of the ocean floor. However, during the winter it moves into shallower areas in sheltered waterways between the islands of British Columbia and Northern Washington, reportedly to lay its eggs. Those that succumb during this migration leave shells within reach of skin divers. Most of the year, the snails live at depths below those usually frequented by divers, but they do get picked up in nets. These are generally otter-trawl nets—so named for the two door-like otter boards that sail out in the water on each side and hold the net open while it is pulled, or trawled, through the water to catch fish.

Shells of the Tabled Neptunes are sturdier than those of the Pribilof Neptunes, but not nearly as thick sided as the similarly shaped shells of the Frilled Dogwinkles of the open coastline. They are slender and of moderate size, occasionally reaching 3 or 4 inches in length. The whorls are rounded up to the part of the whorl called the shoulder. A sharp ridge is then formed and the whorl recurves in a concave manner to the suture. This leaves a rounded groove at the top of the whorl next to the suture so that the shells have a stepped appearance.

The aperture is teardrop-shaped, like the Pribilof's. The nose is fairly long, slightly twisted, and has an open canal. Shell texture is roughened by numerous cords spiraling around the whorls. Growth lines are indistinct, but occasional jagged lines of repaired breaks of a previous lip show. Color is yellowed white, with fairly bright yellow inside the aperture. This bright coloration, unfortunately, fades rapidly in dead shells so is rarely present in the empty shells found by divers.

The TABLED NEPTUNE, *Neptunea tabulata,* on the left, was collected from about 30 feet of water depth near San Juan Island Park, Washington, and contained a hermit crab. A shrimp net scooped up the specimen on the right, off Tillamook Head, Oregon, from a little over 200 feet of water. The jagged line crossing the body whorl above the aperture shows where the shell was previously broken. Newer growth extends beyond and has repaired the break.

CHOCOLATE WHELK

Neptunea smirnia (Dall)

Although occasionally referred to as the Common Northwest Neptune, this is one of the less common and most impressive of our shells. It seems to stay in deep water and is only infrequently taken in otter-trawl nets. This suggests that the range is deeper than that of other Neptune species, starting at around 200 feet and extending beyond. While the author has found a few dead Pribilof Neptunes at 50 feet—and quite a few Tabled Neptune shells at around 80 feet—he has never picked up a Chocolate Whelk in his skin-diving experience.

The shells of the Chocolate Whelk are large and sturdy, proportionately as thick walled as many intertidal species. Even so, some show lines of previously broken lips, probably the result of encounters with predators. There is a variation in relative width of the shells—some are slender, some rather plump.

Although not normally as thick as the Pribilof Neptune, the slender specimens of the Chocolate Whelk become longer, some reaching a length of 6 inches. Whorls are rounded, resulting in distinct sutures, and in very young juveniles a knob often remains on the end of the spire—the nuclear whorl that was formed while the snail was still in the egg.

Unlike other Northwest Neptunes, the Chocolate Whelk is attractively colored. As the name implies, the exterior is smoothly textured and the color of milk chocolate. The aperture, in contrast, is lighter, and at times edged in white. In older specimens, this lighter-colored lip of the aperture is often thicker and flared out to some degree.

Shrimp nets captured these CHOCOLATE WHELKS, *Neptunea smirnia,* west of Newport on the Oregon Coast. The nuclear whorl is still visible at the tip of the spire of the small one.

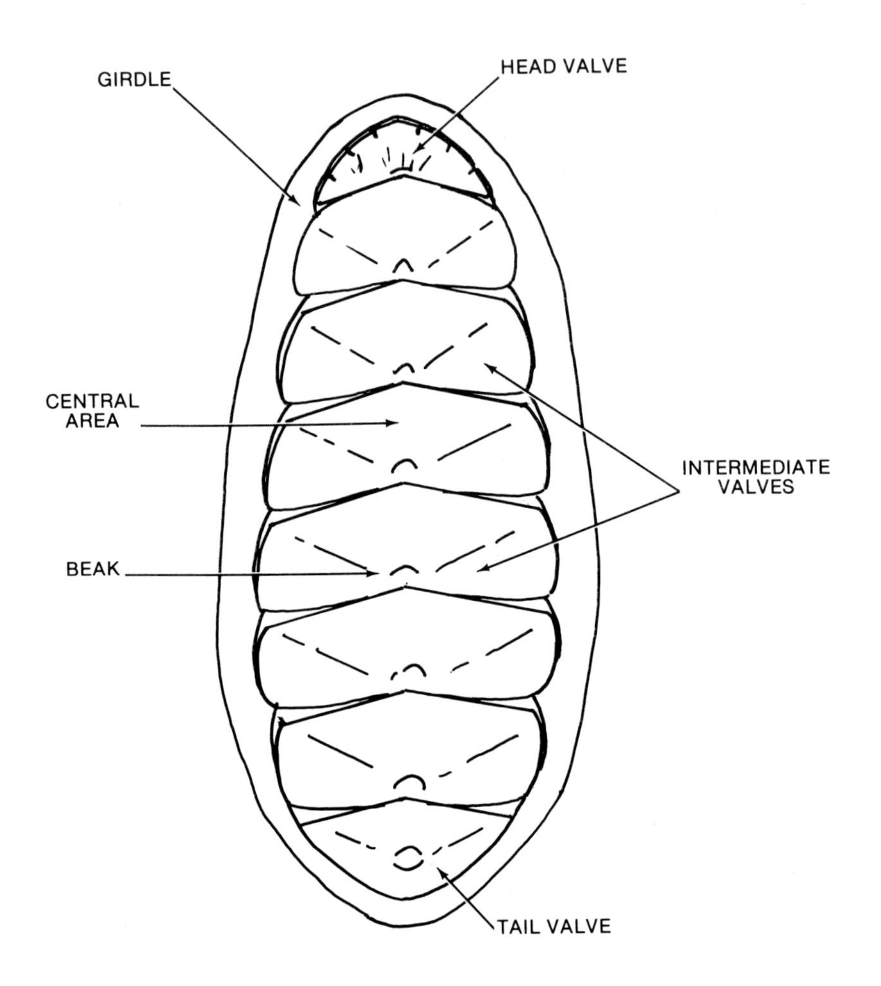

Diagram of Chiton

PART TWO

CHITONS

Like the snails, the chitons move about on a stomach-foot. In many ways they resemble the herbivorous limpets and abalones. Instead of having a single shell for protection, though, they have an overlapping system of eight shelly plates. From this armor system their scientific name is derived. It is Polyplacophora, meaning many (poly) plates (placophora).

The Pacific Northwest has many chitons, or plated mollusks, including some unusually colorful ones, as well as the world's largest plated mollusk, the Giant Chiton. More chitons are found on the Pacific Coast than on the Atlantic. In fact, most intertidal and shallow-water subtidal rocks along our open coast and sheltered waterways contain one or more of the various species. They cling tenaciously to the rocks, their low profile and broad, clinging surface enabling them to survive even in places avoided by most other mollusks.

Chitons are reportedly nocturnal, but they can often be observed moving about rocky surfaces in the daytime, scraping up algae with their rough file-like tongue or radula. Some scientists believe that chitons occasionally feed on animal flesh, but they are normally observed only in areas of seaweed. During resting periods they may seek indentations in rocks to escape predators and wave action.

Protecting the foot and viscera of the chiton, is a tough, gristle-like girdle along the outer edge of the plates. This may seem like an effective defense, but it does yield to such predators as the starfish. Evidence often remains in the form of the eight plates still held together by the indigestible girdle in a shape something like an oval bowl. When this breaks down, the beachcomber will find only the intermediate V-shaped plates and rounded end plates scattered about. Some collectors refer to such intermediate plates as "butterfly shells" and to the end plates, especially the head valve, as "false teeth."

This LINED CHITON, *Tonicella lineata,* seems to be hitchhiking on the shell of a DUSKY TURBAN. Photograph was taken intertidally at Cape Arago, Oregon.

Cleaned and re-assembled chiton plates form patterns in a collection.

Left: Interior view of the plates of a LINED CHITON, *tonicella lineata,* from Seal Rock, Oregon.

Right: Exterior of the re-assembled plates of a RED CHITON, *Tonicella insignis,* collected subtidally in Hood Canal, near Pleasant Harbor, Washington.

RED CHITONS

Chitons in this family are usually red with white stripes, although several variations occasionally occur, including purple-tinged specimens and some that are basically white. The Pacific Northwest has two moderate-sized chitons of this group. Both are fairly common within their range but dwell deeper than several other species; hence they are not so often observed. *Ischnochitonidae*

LINED CHITON

Tonicella lineata (Wood)

The Lined Chiton is small, seldom exceeding an inch in length, but it is one of the most colorful chitons to be found in the rocky areas of Northwest shores. Its red plates are usually decorated with white striping and white centers. (Occasionally the central area of the plates is a contrasting red.) Variations of the striped patterns are also found. The interior of the plates is white with a blush of pink that forms a band down the center of the cleaned and reassembled shells.

The Lined Chiton is primarily an intertidal dweller and, although it normally seeks shelter beneath rocks, shells, or even Sea Urchins, it can often be found at low tide. It dwells both the length of the open coast and in the sheltered waterways.

RED CHITON

Tonicella insignis Reeve

The Red Chiton is the common chiton of this family in British Columbia and northern Washington, which is the southernmost part of its range. It is about the same size as the Lined Chiton (one inch) but darker in color. The Red Chiton is brick-red, with a white-lined pattern finer and most distinct near the overlapped parts of the plates. Outer portions of the plates seem to be without pattern.

The Red Chiton lives deeper than the Lined Chiton and is abundant on subtidal rocks in some parts of Hood Canal, Washington. Occasionally there appears to be an overpopulation, with numerous tiny, perhaps stunted, Red Chitons being found on isolated objects like bottles.

Plates of a GIANT CHITON, *Cryptochiton stelleri,* collected from the base of subtidal rocks in Whale Cove, Oregon.

GARMENTED CHITONS

This group of chitons, whose family name means spiny garment, completely covers its shelly plates under a rough and durable mantle. Because of this, they are often mistaken for some other kind of marine life. Within the Pacific Northwest we have only one representative of the group. *Acanthochitonidae*

GIANT CHITON

Cryptochiton stelleri (Middendorf)

The Giant or Gumboot Chiton does not resemble other chitons; no plates are visible on the living animal. It is a huge oval creature, sometimes over a foot in length, shaped roughly like the upper half of a loaf of bread. The dark-red color and rough surface of the skin, though, are more similar to a brick. Unlike chitons that remain on rocks, the Gumboot (the underside is about the color and texture of a gum-rubber boot) is at home on the sand or mud. During low tides it often leaves the rocks and seeks the depths of large tidepools. It can be discovered at the base of intertidal rocks, hiding under the edge of a boulder, or in tidepools. Subtidally, Giant Chitons are more common; they are frequently seen by skin divers.

Beneath the thick skin of the Giant Chiton are the typical eight plates of the polyplacophorans. In this case, though, color is almost totally absent. With the exception of an occasional set with a blush of pink, Giant Chiton plates are a polished white. Those in the middle, called intermediate valves, resemble white butterflies if found individually. The shiny head valve is the chiton plate that most closely resembles an upper denture complete with white teeth.

Of all our Northwest chitons, the Gumboot, because of its impressive size and limited availability, is most desired by collectors. The plates are well earned, not only from the standpoint of finding, but also in terms of labor involved in cleaning. Even then, one is apt to encounter broken plates, particularly in specimens collected on the open coast, for the Gumboot does not cling to the rocks consistently and may have been tossed about in a storm surf.

A MOSSY CHITON, *Mopalia muscosa,* feeds on algae at Cape Arago, Oregon. Note how the bristly girdle blends with the algae in this intertidal location.

Many collectors prefer to leave the chiton girdle, soaked in glycerin, attached to the plates.

Left: Prepared specimen of MOSSY CHITON, *Mopalia muscosa.*

Center: Interior view of MOSSY CHITON plates.

Right: Prepared specimen of HAIRY CHITON, *Mopalia ciliata.*

A GUMBOOT (GIANT) CHITON, *Cryptochiton stelleri,* leaves the rocks and looks for a tidepool to hide in during an extremely low tide at Yaquina Head, Oregon.

Left: Two specimens of WOODY CHITON, *Mopalia lignosa,* collected intertidally north of Oceanside, Oregon. The one closest to the middle is more contrasting than usual.

Right: Two specimens of the highly variable SWAN'S CHITON, *Mopalia swanii,* from the Strait of Juan de Fuca.

FUZZY CHITONS

This family of chitons includes species that are among the most common of the intertidal group. They are characterized by rather flattened bodies and plates, with wide girdles bearing some form of short hair or fur. The dominant color is dark green, though individual specimens may be yellow or orange. Because fuzzy chitons have the ability to clamp tightly to rocks and withstand the surf, they may be found during moderately low tides. *Mopaliidae*

MOSSY CHITON
Mopalia muscosa Gould

The Mossy Chiton is one of the most common inhabitants of intertidal and subtidal rocks and is found the length of the Pacific Northwest Coast. It often is 2 inches long and may reach 4. Plate colors of most Mossy Chitons are variations of dark brownish-green, but older, larger specimens are usually worn to a lighter green layer. The girdle surrounding the oval set of plates is fairly wide and covered with closely set bristly hairs. These hairs blend with the fine algae found on the rocks where the Mossy Chiton reside, aiding in camouflage.

Within, the cleaned plates are white, tinted with blue-green. The green is most prominent in the center and is considered an identifying characteristic.

HAIRY CHITON
Mopalia ciliata (Sowerby)

At first glance this would seem to be a repeat of *Mopalia muscosa*, for both have fairly broad girdles studded with bristles or short hairs. The Hairy Chiton is smaller, usually not over 2 inches long, and can be discerned by the more heavily textured plates. Diagonal ridges form V patterns in the center of each intermediate valve, while the sides of the valves tend to be granular. Colors are much more variable and may include yellows, oranges, and reds, as well as green and brown. The interior of the plates is paler than that of the Mossy Chiton, usually white, tinted only lightly with green, orange, or pink (consistent with the dominant exterior color).

WOODY CHITON

Mopalia lignosa Gould

Like the Mossy and the Hairy chitons, the Woody Chiton is found throughout the Northwest coastline on intertidal and shallow-water subtidal rocks. It, too, is about 2 inches in length and has a hairy girdle, although not as bristly as the Mossy Chiton. The Woody Chiton is generally discernible from its fellow Mopalia by the striking pattern of lightning stripes that are found lengthwise to the valves. Exterior colors may be any of several: olive green with darker stripes, white markings on dark green or black, occasionally white on brown or orange. The striping, though faint on some specimens, is generally present and may be considered a characteristic. The interior of the plates is white, tinted with green.

This chiton is a favorite with collectors because of the distinctive and attractive pattern.

SWAN'S CHITON

Mopalia swanii Carpenter

Though Swan's Chiton is much less common than the Woody, Mossy, or Hairy chitons, it is often found in Northern Washington and British Columbia. A large percentage of this species has colorful markings, which makes them highly sought after. The color variations include yellow, orange, olive green, and brick red. Patterns are speckled or striped—or individual plates may vary so much from others on the same chiton as to seem mismatched. Interiors of plates are apt to be ivory, tinted with the dominant exterior color. Actually, it is the girdle that is the most consistent part of the Swan's Chiton for identification. Unlike the girdle of its fellow Mopalia, the Swan's Chiton girdle is soft and velvety, the ciliation a mere fuzz.

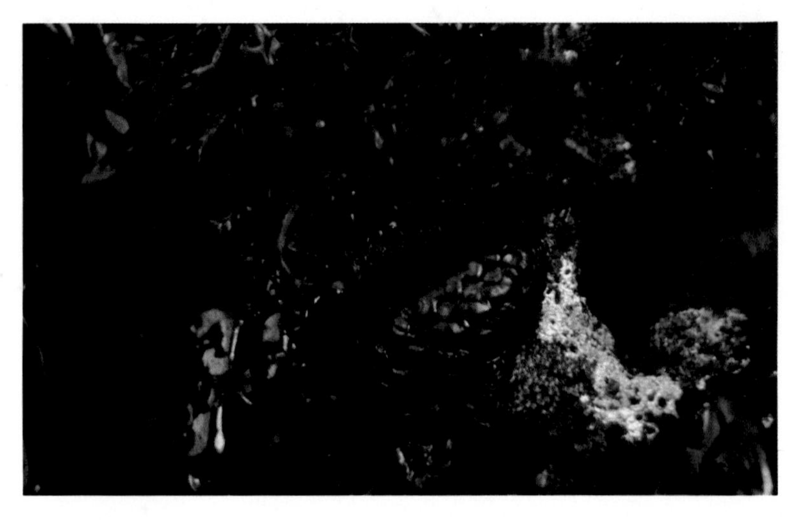

This BLACK LEATHER CHITON, *Katherina tunicata,* is difficult to see among the dark seaweed and basalt rock in Whale Cove, Oregon.

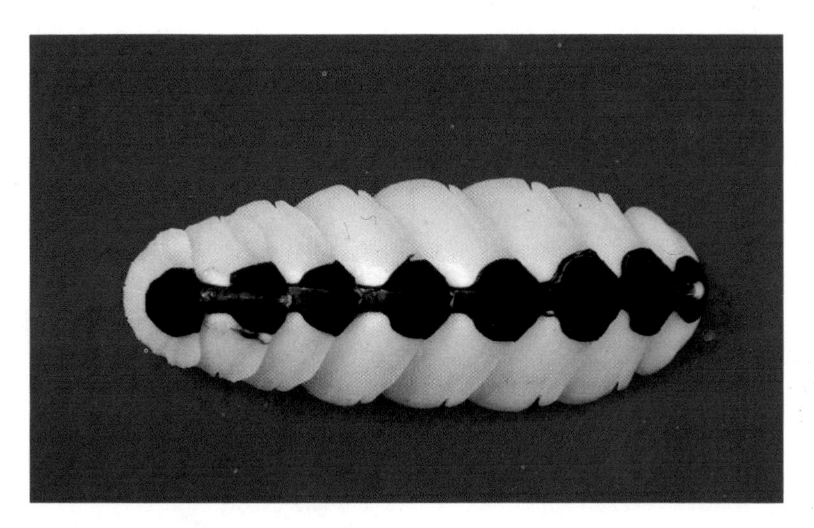

A re-assembled set of plates of the BLACK KATY shows how only the black central area is normally exposed. This specimen, also collected at Whale Cove, was fairly old, so portions of the central areas are worn away.

BLACK LEATHER CHITON
Katherina tunicata (Wood)

The Black Leather Chiton, or Black Katy, bears little resemblance to the Mossy, Hairy, or other Fuzzy Chitons, yet is considered to be in the same family. Its girdle is extremely large, covering most of the plate area, and, in addition, is smooth and hairless, with a texture something like wet leather. This black girdle, however, and black exposed portion of the plates make it difficult to find among the seaweeds and crevices of the dark basalt reefs and headlands so common to the Northwest.

Larger than its relatives, with the exception of the largest specimens of the Mossy Chiton, the Black Katy often reaches a length of over 3 inches. It is also deeper; that is, its plates are more V-shaped, allowing for a larger mass of body tissue within the area protected by the plates. Living fairly high in the intertidal zone, the Black Leather Chiton can generally be observed on most low tides. However, colonies of them are found in some rocky locations, while other parts of the coast may be devoid of Black Katies. The author has found them most abundant subtidally in sheltered coves of the open coast, places like Whale Cove, Oregon, where there is a large submerged area of black basalt.

The appearance of the cleaned plates of the Black Leather Chiton is unlike that of any other Northwest chitons. They seem to be made of two materials: a blob of black material topping glistening white plates, which, in the intermediate valves, resemble blade-like wings that extend out and down from the black center. This black blob is the only portion exposed by the mantle while the animal is alive. At times, though, the black part becomes worn away, breaking the Katy's protective coloration with a series of dull white spots. All the rest of the plate material (all the interior) is white.

Within its home range, the Black Leather Chiton feeds upon the algae on the rocks; it is, in turn, fed upon by the starfish which, though preferring bivalves, occasionally eat chitons. Evidence of this is apparent at the base of subtidal rocks where starfish have been feeding. There, scattered about like many ceramic birds with extended wings, will be the discarded plates of consumed Black Katies.

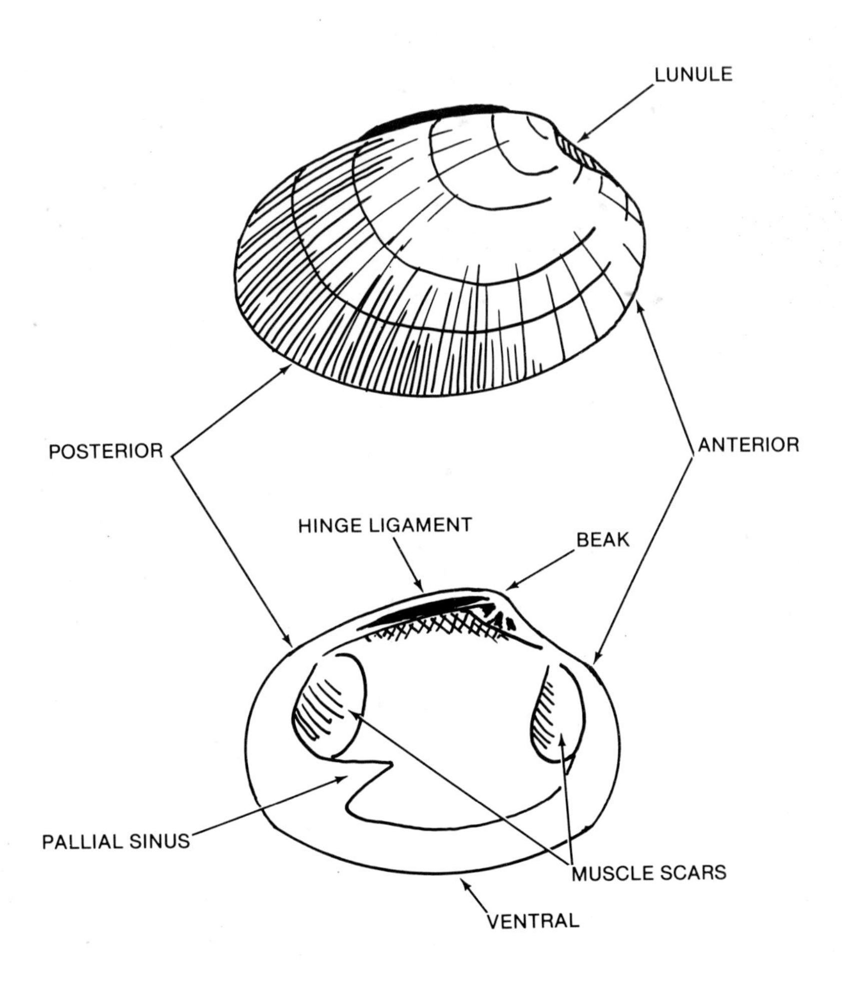

Diagram of Clam

PART THREE

CLAMS
and Other Bivalves

All mollusks serve as food for other animals; most are food for mankind. In this country, with the exception of the abalones, the edibility of the snails is largely ignored. On the other hand, all of the bivalves described in this text are edible, and most of them are regularly consumed by human beings.

While the shells of the large Gooeyduck and Gaper clams are impressive, these clams are eagerly sought for their eating quality, with most shells being discarded. Even shells of the shiny brown Razor Clam are quickly cast aside, forgotten in preference for their tasty flesh. At the other extreme, the delicious adductor muscles of Pink Pacific Scallops are frequently wasted, ignored by collectors seeking the colorful shells.

Like the snails and other gastropods such as nudibranchs or shell-less snails, the bivalves, or pelecypods, are soft-bodied mollusks. Most differ from snails by leading a stationary life; indeed many, like oysters, have lost the power of locomotion in adult life. All are filter-feeders, drawing to them the minute plankton upon which they dine, and straining it from the water. By utilizing such free-floating material for food, the bivalves have no need to move. Movement during the larval stage accomplishes distribution of new populations.

A few bivalves, such as swimming scallops and digging clams, use locomotion as a means of defense—the scallops by jetting water, the clams by digging with their "hatchet foot" (pelecy / pod). Most bivalves depend on the strength of their two shells and/or burrow for protection. A few, like the Pearly Monia and the Rock Scallop, use camouflage as well.

Among the bivalve population of the Pacific Northwest are a number of colorful shells. Most notable are the scallops, but attractive shells are also found among clams. The Rose-Petal Semele, the Sunset Shell, and the Pink Macoma are prime examples.

Top: PINK SCALLOP, *Chlamys hericia,* collected from a water depth of 20 feet at San Juan Island, Washington. Note the byssal notch in the lower (lighter-colored) valve.

Bottom: HINDS' SCALLOP, *Chlamys rubidus,* was washed up on the beach near Port Ludlow, Washington.

Shown are both valves of the WEATHERVANE SCALLOP, *Pecten caurinus,* with lower valve toward the bottom of the illustration. Specimen was taken in nets from deep water west of Astoria, Oregon.

SCALLOPS

Scallops have long been the most sought-after bivalve shells. Their shape is well known in art forms, both ancient and modern. Among bivalves, scallops have the most rapid form of mobility, accomplished by jetting water from between their valves. They also have two rows of tiny, but well-developed, eyes in the mantle tissue, one along each edge of their shells. *Pectinidae*

PINK SCALLOP

Chlamys hericia (Gould)

The Pink Scallop is the Northwest bivalve shell most desired by collectors; it is outstanding for its graceful shape and delicate colors. The Pink Scallop is found at various points along the Northwest coast, but is most abundant in the sheltered waterways of northern Washington and British Columbia. It prefers subtidal rocky areas where it attaches by a byssal strand and remains unless disturbed, its shells agape, its striped black- or orange-and-ivory mantle studded with two rows of tiny, glistening blue eyes. While alive it is normally coated with a layer of yellow sponge which hides the shells. Many people gather Pink Scallops for shell collections but they are also harvested for the tasty adductor muscles.

The shells of the Pink Scallop are rounded on three sides, pointed at the hinge. Extensions, referred to as ears, are located on each side of the point (beak) and form a straight line at the hinge. The byssus passes through a byssal notch located where one ear forms an angle with the principal part of the shell. Only the lower valve (on which the scallop rests) has a byssal notch.

Color of the shell is basically white with bright pink stripes radiating out from the beak. The stripes are lighter on the lower valve, at times apparent only at growth rings. The texture consists of radiating rough ridges which are also apparent on the inside of the shell. Shells are commonly 2 to 3 inches across the widest point. They are shallowly concave, presenting a very slender side view.

HINDS' SCALLOP

Chlamys rubidus (Hinds)

The Hinds' Scallop shares the rocky subtidal habitat of the Pink Scallop. About the same size (around 2 inches across) and generally covered with the same yellow sponge, it is difficult to distinguish from the Pink. However, the cleaned shells of the Hinds' Scallop are noticeably different.

Compared to the Pink Scallop, the Hinds' shells are more nearly round, smoother with finer radiating ridges, and without the radiating bands of color. The shells are a rosy beige, rather evenly shaded, but brighter at the beak. Though the Hinds' Scallop is less common, it lacks the contrasting color pattern of the Pink and is thus less sought by collectors.

WEATHERVANE SCALLOP

Pecten caurinus Gould

The Weathervane Scallop is the largest swimming scallop in the Pacific Northwest, at times reaching a size of 7 inches across the shell. It lives deeper than its smaller relatives and is rarely seen alive even by skin divers. The most common method of obtaining Weathervane Scallops is in trawl nets where they are picked up incidentally with bottom-dwelling fish. Unfortunately, the nets usually chip the edges of the Weathervane's fragile shells.

The valves of the Weathervane Scallop are round, have obvious ears and a byssal notch in the lower valve. Both valves are shallowly concave but have a number of differences. The lower valve has a pattern of approximately 16 flattened ridges radiating from the beak—less apparent at the edges adjoining the ears or wings. Ridges on the upper valve are often less distinct but not flattened. Lower valves are light beige, darker on the ridges and concentric growth lines; upper valves are a reddish mahogany brown with ivory ears. Tiny borers infest the upper valves of adult Weathervane Scallops, piercing them with numerous black pinholes.

PURPLE-HINGED ROCK SCALLOP

Hinnites giganteus (Gray)

In adult life the Purple-hinged Rock Scallop appears unrelated to its swimming relatives. The orange mantle seems typically scallop, but the rows of eyes stare out from what appears more like a worn and encrusted rock than a delicate and colorful shell. Tiny Purple-hinged Rock Scallops start their existence like the other Pectinidae of the Pacific Northwest, swimming about when disturbed, sporting graceful, often colorful shells. But, by the time they are adults, they have lost their mobility and cemented the edge of their lower valve to a rock. From this fixed position they grow out, increasing the attached area like an oyster. During this stage the shells become thickly encrusted and riddled by boring sponges. Such older scallops may be 9 inches across and weigh several pounds (shells and internal parts), and are the source of a tasty meal for the diver who pries the scallop from its resting place.

To the shell collector, the juvenile- and intermediate-stage shells are most interesting. In the juvenile stage (up to about one inch across) the shells resemble other swimming scallops with their oval outlines and ears at the hinge. During this time the shells are bright yellow, orange, ivory, or brown. The interior is ivory and, even at this size, has a hint of purple at the hinge. At the end of the juvenile period, the edge of the lower valve is cemented to a hard surface. Progressing into intermediate growth, the shells become thicker, somewhat irregular, with radiating ridges of flat, curving spines. They are usually chocolate brown outside and off-white inside, with an enlarging area of purple at the hinge. During this time the shells collect camouflaging algae and tubeworms.

Purple-hinged Rock Scallops are found the length of the Pacific Northwest coastline, on subtidal reefs offshore, and in the sheltered waterways. Most reside at the 20- to 30-foot depth range. Only the borer-riddled upper valve, not being cemented down, washes to the beach. Divers, however, find this scallop by spotting the line of the orange mantle.

PURPLE-HINGED ROCK SCALLOPS, *Hinnites giganteus,* collected in the intermediate or young adult stage, from Haystack Rock near Pacific City, Oregon. The smoother orange-yellow area is the part of the shell that developed during the juvenile (free-swimming) period.

SEMELES

Semele clams have rounded, frequently colorful shells of moderate size. Named after a Greek mythological figure, they are among the most attractive bivalves. Most Semeles are confined to warm seas, with only one species living in the Pacific Northwest. *Semelidae*

ROSE-PETAL SEMELE
Semele rubropicta Dall

This relatively small clam (about 2 inches in length) seems to prefer beaches of clean gravel in the sheltered waterways of northern Washington and British Columbia. Most specimens of which the author has been aware have also come from just below the intertidal area. The shape of the Rose-Petal Semele shell is rounded oval, although there is a fairly prominent beak placed toward the siphonal end. A slight furrow near the posterior margin of one valve causes a wrinkle to be carried into the suture between the valves. Color is the shell's most distinguishing feature. Rays of bright pink or red radiate out from the beak of an otherwise white shell. Because this color fades if the shell is not kept out of the sun, it may be lacking in beach-worn specimens.

GARI SHELLS

This is another group of predominantly warm-water clams which has one species inhabiting the Pacific Northwest. Shells are usually colorful, elongated, and moderate in size. *Sanguinolariidae*

SUNSET SHELL
Gari californica (Conrad)

The Sunset Shell Clam is distributed in the Hood Canal and Puget Sound regions, where divers find it at about an 80-foot depth. It is also occasionally discovered just offshore from Alaska southward. The shells of this clam are elongated and thin, suggestive of the Razor Clam. It also has a similar external hinge. Length may reach a little over 3 inches. Exterior colors are light beige with darker, rosier stripes radiating from the beak.

MACOMAS

Members of the family of Tellins, the Macomas are significant in the Pacific Northwest both from the standpoint of color and for their contribution as edible clams. Although tasty, they are usually ignored in favor of larger species. *Tellinidae*

IRUS MACOMA

Macoma irus (Broderip & Sowerby)

This moderate-sized clam, with a length of up to 2 inches, is not plentiful, so seldom becomes a clam digger's meal. It is, however, occasionally encountered over a wide range of clam flats, where it is accidentally uncovered along with more populous clams.

Like all Macomas, the Irus Macoma is tapered at the siphonal end and rounded at the anterior. The valves are of equal size, with little distortion, although there is a tendency for the ridge running along the pointed end of the shell to cause an indentation where it meets the ventral edge. Shell colors vary from white with a blush of yellow to dull orange. Concentric growth rings are often highlighted with darker bands. A darker periostracum, present along the edges, is usually worn off the major part of the shell.

BALTHIC MACOMA

Macoma balthica (Linne)

The Balthic, or Pink Macoma, has a small, fragile shell (less than an inch in length) that is frequently found along the edges of our estuaries and sheltered coves. When the empty shells are discovered, they often lie open in a butterfly fashion, a position allowed by the external hinge ligament found in Macomas. Such bright little shells are widespread, for the Balthic is not an exclusively Northwest resident; it is found around the polar regions and into the Atlantic.

The shell is egg shaped, rounded at the anterior end and tapered at the siphonal. It is smooth on both the interior and exterior. Its colors may be a variety of pink and rosy hues, at times with concentric bands of white and pink.

The Pacific Northwest shores contain a few colorful clams.

Top: ROSE PETAL SEMELE, *Semele rubropicta,* raked from shallow-water gravel near Pleasant Harbor, Washington.

Bottom: SUNSET SHELL, *Gari californica,* found at a depth of 80 feet near Hoodsport, Washington.

Top: IRUS MACOMA, *Macoma irus,* is sometimes colorful, like this one found on the tide flats near the Marine Science Center, Newport, Oregon.
Bottom: BALTHIC MACOMAS, *Macoma balthica,* collected at the same location, are typical.

BENT-NOSE CLAM

Macoma nasuta (Conrad)

The Bent-nose is a moderate-sized, white-shelled clam about 2 inches in length, found in many estuaries along the Pacific coast. Since it is able to tolerate a wide range of salinities, populations of them are found well up into estuary systems. Although not widely used for food now, it was consumed by local residents in earlier times when transportation was less efficient.

The shell outline of the Bent-nose Clam is egg-shaped like most other Macomas. A distinguishing difference, though, is the tapered, siphonal end, which is bent to one side. This presumably aids the Bent-nose when lying on its side in the mud or gravel with its split siphons reaching up to the water. The siphons of the Macomas are divided into two tubes, one for the in-current, one for the ex-current flows.

WHITE SANDCLAM

Macoma secta (Conrad)

The White Sandclam is the largest of our Macomas, at times growing to a length of 4 inches. Its shell also has a proportionately greater height and is stubbier than its relatives. Unlike the Bent-nose, which lives well up the estuaries, and the mud-dwelling Pink and Irus Macomas, the White Sandclam prefers a sandy environment near the mouth of bays and sheltered beaches. While it is somewhat widely distributed south of Vancouver Island, it is neither abundant nor easy to find, so is not commonly used as a food supply.

The shell of the White Sandclam is thin and fragile, particularly in areas of newer growth at the edges. At the tapered edge is a thin flap of shell which projects beyond the ridge usually found on Macomas. This is most apt to be broken away from shells found on the beach. The shell is pure white on both the exterior and interior surfaces. A thin, transparent periostracum normally remains on the exterior, giving the shell a shiny appearance. The black, external hinge is in sharp contrast.

Top: WHITE SANDCLAM, *Macoma secta,* collected near the mouth of Yaquina Bay, Oregon.
Bottom: BENT-NOSE CLAM, *Macoma nasuta* (left), dug in the upper Yaquina estuary. INCONSPICUOUS MACOMAS, *Macoma inconspicua,* were taken in shrimp nets west of Newport.

INCONSPICUOUS MACOMA

Macoma inconspicua (Broderip & Sowerby)

Like the Pink Macoma, the Inconspicuous Macoma is small and delicate, generally less than a half inch in length. It has, however, rather dull white shells, often so thin that they are translucent. A thin, beige periostracum gives the shells some color.

The Inconspicuous Macoma is not of significance as far as human diet is concerned. However, it apparently is abundant in the deeper waters offshore and is frequently found in the stomachs of deepwater fish. In the sheltered waterways of British Columbia and northern Washington, as well as offshore, the Inconspicuous Macoma is a regular part of the diet of large Twenty-rayed Starfish (*Pycnopodia helianthoides*). The author has sometimes found a dozen or more such clams in the stomach cavity of this starfish.

COCKLES OR HEART-CLAMS

This family contains many species of roughly heart-shaped clams. Its world-wide representatives include the Egg Cockles and many species well decorated with variations of ribbing. They are regularly used for food in other lands, but are generally ignored in the Pacific Northwest except as bait clams. *Cardiidae*

BASKET COCKLE
Clinocardium nuttallii (Conrad)

The Basket Cockle is a common clam throughout our estuaries and sheltered waterways. In subtidal waters it can be gathered from the surface of the muddy bottom, and intertidally may be raked from clumps of seaweed. Although used primarily for bait, its flesh makes excellent chowder.

Shells of the Basket Cockle are rounded, deeply concave, and have very prominent beaks. Adult specimens become flattened and longer on the posterior end. Their distinguishing texture consists of 37 large, rounded and somewhat file-like ridges radiating from the beak to the ventral edge of the shell, where they form a toothed pattern. Frequent concentric lines that give a file-like roughness to the ribs also form somewhat of a basket pattern. Color varies, but tends toward light beige in Oregon estuaries, brown in Hood Canal. Length may reach nearly 4 inches.

FUCAN COCKLE
Clinocardium fucanum (Dall)

The Fucan Cockle is less common and a deeper water dweller. Its range is said to be throughout the Pacific Northwest, but appears to be most common in waters adjoining the Strait of Juan de Fuca. Generally, the Fucan Cockle is smaller, about 2 inches in length, more oval, and can be distinguished from its common relative by the number of ridges, about 50, on the shell. Color is also a darker brown.

Top: BASKET COCKLE, *Clinocardium nuttallii,* raked from the tide flats of Yaquina Bay, Oregon.
Bottom: FUCAN COCKLE, *Clinocardium fucanum,* collected from about 30 feet of water near Seabeck, Washton.

VENUS CLAMS

This is one of our largest and best known families of clams. The Butter Clam and several species of Littleneck Clams, familiar to both sport and commercial clam digger, are within this group. The shells of Venus Clams are known for their oval symmetry, and often for their sculptured texture. *Veneridae*

SMOOTH WASHINGTON (BUTTER) CLAM

Saxidomus giganteus Deshayes

The Butter Clam is the largest of the Venus Clams found in the Pacific Northwest. During exceptionally low tides, people flock to Butter Clam beaches with shovel and bucket to gather their favorite clams. Where the Butter Clams are plentiful, Gapers and other species are apt to be left behind by the diggers who seek only the Butters to fill their limits. In some places they are dug commercially as well.

Smooth Washington or Butter Clams may be found anywhere from intertidally to depths of about 30 feet. Subtidally, they tend to live closer to the top of the mud where they are more within reach than some of their neighbors, so form a regular part of the diet of starfish. In Tillamook Bay, Oregon, the author has found the Butter Clams to be about 6 inches deep, just below the Bent-nose and Littleneck Clams, but several inches shallower than the Gapers. Intertidally, where the clams tend to dig deeper for protection, one may have to dig nearly a foot to find the Butter Clams.

What we call the Butter Clam in the Northwest is not necessarily the same species as the "Butter Clams" of other areas. In California, a similar species of Butter Clam, also called the Washington Clam (not Smooth Washington), predominates. Shells of both the Washington and Smooth Washington Clams are fairly thick, widely oval in shape, and fairly deeply concave. Both ends are rounded, and the beak is close to the foot (anterior end). Shells of both these Butter Clams are grayish on the outside, more nearly white inside. The southern species, however, has a series of ridges like encircling growth lines, while the Smooth Washington has only a few fainter lines of growth. Both become about 5 inches in length.

SMOOTH WASHINGTON (BUTTER) CLAM, *Saxidomus giganteus,* picked up in about 20 feet of water near Seal Rock Park, Hood Canal, Washington.

KENNERLY'S VENUS

Humilaria kennerlyi (Reeve)

The Kennerly's Venus Clam is one of the most interesting clams in the Pacific Northwest both from the standpoint of its sculpture and from its tenacity. It is fairly common throughout the waterways of Washington and British Columbia, occurring from 20 to 80 feet in depth. In this environment it is constantly subjected to onslaughts of starfish that dig them out of the silt and gravel, where they live fairly close to the surface. However, this clam is very strong—so much so that ridges of shell break off before the clam gives in to the predator. The author has seen large starfish moving away from chipped, but very much alive, Kennerly's Venus clams.

The shells of the Kennerly's Venus are roughly oval, with the beak far toward the anterior end. Shells are deeply concave, thick-walled, and fitted tightly together. Color is a dull white, usually tinged with gray or brown, depending upon locale. They have a unique sculpture, the main part of which is the series of concentric ridges that cover the exterior. These are flat sided with V-shaped grooves between, and of a fairly brittle material. It is the breaking away of these ridges, giving little for the starfish to grasp, that aids substantially in the Kennerly's defense.

Another aspect of this shell's sculpture is the portion just anterior to the beaks. There, outlined from the other sculpture is an area of finer lines, which, when the two valves are joined, form a pattern roughly like a valentine heart. This pattern is called a lunule. Many clams lack this feature while on others it is prominent, thus its presence may be used in scientific keys as an aid to identification.

The Kennerly's Venus is a fairly large clam, at times becoming close to 4 inches in length. If it were not for its deep habitat, it might well be as popular an edible clam as its close relative the Butter Clam.

KENNERLY'S VENUS, *Humilaria kennerlyi,* collected at a depth of 80 feet at White Rock, San Juan Islands, Washington. The clam was alive but in an area recently disturbed by a starfish. Note the places where the encircling ridges have been chipped.

PACIFIC LITTLENECK

Venerupis staminea (Conrad)

Although among our smaller Venus clams, Pacific Littlenecks are regarded as a delicacy, most frequently consumed as "steamers." They are found intertidally both in estuaries and on the open coast, usually in gravelly, somewhat sheltered beaches. Gathering them is accomplished by raking intertidal gravel, or simply by picking them up subtidally.

The Pacific (or Native) Littleneck is a stubby clam with almost oval shells, averaging about 2 inches in length. There are a number of variations in texture of the shell, which, with other factors, have caused some experts to divide the identification of the Pacific Littleneck into several subspecies. However, it usually consists of numerous fine threads radiating from the beak, crossed by encircling fine ridges and growth lines. Colors are also variable but tend toward a yellowed white along the Oregon Coast, brown in some parts of Hood Canal, Washington, and cream with brown zigzag markings in the islands along the Strait of Georgia in British Columbia.

JAPANESE LITTLENECK

Tapes philippinarium (Adams & Reeve)

Also known as the Manila Clam, this species was introduced into Puget Sound and has been gradually spreading into adjoining areas. It is now being introduced into some Oregon estuaries. Where it has become established, it is sufficiently abundant to be harvested commercially. Most "steamer" clams sold in our markets and restaurants are this species.

The shell of the Manila Clam differs from the Pacific Littleneck by being more elongated, tapering at the anterior end before rounding, and being more colorful than most Pacific Littleneck shells. Basic exterior colors are beige with mottlings of light brown, though in some areas, darker colors are usual. Interior colors are white with a distinguishing shading of purple at the posterior or siphonal end. Texture, similar in pattern to the Pacific Littleneck, is more distinct at the ends, fading out in the center.

THIN-SHELLED LITTLENECK

Calithaca tenerrima (Carpenter)

Rarest of our Littleneck clams, the Thin-shelled Littleneck is the largest and most interesting in terms of its shell. Although reportedly found all along the Pacific Coast, it is encountered most often in sheltered waterways like Puget Sound and Hood Canal. Since it dwells deeper than the other Littlenecks, skin divers are more apt to find it.

The Thin-shelled Littleneck is oval in outline, similar to a typical Venus Clam, but the edges of the shells are straight instead of tightly cupped-in like their relatives. As the name implies, they are also thin and fragile. Were they to live in the surf zone like other Littlenecks, they could not survive. The color of the shells, which may be up to 4 inches in length, is a yellowed white. Most interesting is the texture, which, in addition to faint radiating threads and crossing concentric lines, has concentric delicate ridges, spaced roughly every eighth of an inch.

LUCINES

This family is composed of clams whose characteristics include rounded, rather compressed (flat) shells that are usually well formed, white, and attractive. *Lucinidae*

RINGED LUCINE

Lucinoma annulata (Reeve)

The Ringed Lucine is occasionally confused with the Thin-shelled Littleneck as both tend to live at the same depth, are about the same size, are both uncommon and have similar sculpture of delicate concentric ridges. The Lucine, however, is nearly round in outline, has a thicker shell, and is normally covered with an ivory periostracum. In living specimens the periostracum gives the shells a shiny appearance. The beak is also close to the center of the dorsal edge of the shell.

The Ringed Lucine is generally considered a dweller of moderately deep water. However, in Netarts Bay, Oregon, it is occasionally found in the intertidal zone.

Top: PACIFIC LITTLENECK CLAMS, *Venerupis staminea,* raked from intertidal sand at Thetis Island, British Columbia.
Bottom: JAPANESE LITTLENECK CLAMS, *Tapes philippinarium,* gathered intertidally near Seabeck, Washington.

Top: THIN-SHELLED LITTLENECK, *Calithaca tenerrima,* picked up in about 10 feet of water at Gig Harbor, Washington.

Bottom: RINGED LUCINE, *Lucinoma annulata,* collected intertidally on an extremely low tide at Netarts Bay, Oregon.

RAZOR CLAMS

Many residents of the Pacific Northwest feel that the Razor Clam is something unique to this area. While the Pacific Razor Clam is one of the largest, the Razor Clam family has members in nearly every sea. Generally, they have thin, elongated shells covered with periostacum, and live in the sand of shallow coastal waters. *Solenidae*

PACIFIC RAZOR CLAM

Siliqua patula (Dixon)

This Razor Clam is eagerly sought after by clam diggers all along the beaches of northern Oregon and Washington, particularly during the extremely low spring tides. Razors also occur on beaches of British Columbia and southern Oregon, but are less abundant or less accessible there. Razor Clams are undoubtedly the most popular clam for eating and are commercially harvested. They are a challenge to gather because of their ability to dig rapidly and bury themselves in wet sand. Movement is accomplished by extending the foot, anchoring it by inflating its tissue with fluid, then pulling itself through the sand.

The shells of the Pacific Razor Clam are elongated and rounded at both ends. The dorsal and ventral edges also have a slight curvature making the outline an elongated oval. This differs from most Razor Clams which usually have almost parallel dorsal and ventral lines. The beaks of the Pacific Razor Clams are small and near the hinge. Near the beaks, on the interior of these thin shells, are single reinforcing ribs that broaden and angle as they progress toward the ventral edge.

Basic shell color of the Pacific Razor Clam is white. In living specimens, however, the inside of the white shell is flushed with purple, and the exterior is covered with a shiny olive-brown periostracum. In the periostracum, periods of growth can be seen in dark and light bands of brown. Shell length of the Pacific Razor Clam is regularly around 5 inches, occasionally 6 or more. This is several times that of most of its relatives.

PACIFIC RAZOR CLAM, *Siliqua patula,* dug near Seaside, Oregon, during an extremely low tide.

BLUNT RAZOR CLAM, *Solen sicarius,* found at Port Townsend, Washington, at a water depth of about 20 feet.

BLUNT RAZOR CLAM
Solen sicarius Gould

Although the Blunt Razor Clam is much smaller than the Pacific, its 3-inch size is still larger than most members of the Razor Clam family. These Razor Clams, generally ignored by diggers in the Pacific Northwest, might be harvested if found elsewhere. The Blunt Razor is much less abundant than the Pacific Razor Clam, but it does have the adaptability to live in sheltered waterways while the Pacific Razor is confined to the open coast.

The shape of the Blunt Razor Clam is typical of the family, with the dorsal and ventral sides gently curving but parallel. Both the indistinct beaks and the hinge are at the foot (anterior) end. The siphonal end of the shell is rounded, but the foot end is "blunt," actually appearing chopped off. Both anterior and posterior ends gape widely, giving the assembled shells a tube-like appearance. Colors are basically white, with a blush of pink or brown in the interior. Shells of living clams are covered with a dark brown periostracum which quickly dries and peels off.

EASTERN SOFT-SHELLED CLAM, *Mya arenaria,* dug from the tide flats along the south side of Tillamook Bay, Oregon.

SOFT-SHELLED CLAMS

Like the Gooeyduck and the Gaper clams, the shells of the Soft-shelled Clams do not completely close. They gape open around the siphon. The most consistent identifying characteristic of the Soft-shelled, however, is at the hinge. Here the two shells are not reverse images; one shell (considered the left valve) has a spoon-like projection, called a chondrophore, that holds the hinge ligament. The opposite valve is recessed at the corresponding point. *Myacidae*

EASTERN SOFT-SHELLED CLAM
Mya arenaria Linne

It was originally thought that this clam was introduced from the Atlantic Coast, but its wide range, from Alaska to California, and its presence in Indian shell middens (garbage heaps) suggest that it is native to both coasts. On the Atlantic Coast it is an important commercial clam, but in the Pacific Northwest it is overshadowed by the abundance of larger and tastier clams.

Soft-shelled Clams, however, are found in places where other clam populations do not survive. They have an ability to withstand substantial changes in salinity, making it possible for them to live well up into estuary systems. There, clam diggers are able to locate clam beds by the spurts of water from disturbed clams, or by the figure-of-eight-shaped holes left when the split siphons are retracted.

As the name implies, the shell of the Soft-shelled Clam is relatively thin, but it is no thinner than some of the Macoma species. The shape is an elongated oval, narrower at the siphonal end, giving the shell an egg-shaped outline. Texture of the shell is smooth except for faint concentric lines of growth. On the interior, the pallial sinus is large, extending back over half the length of the shell. The most significant identifying characteristic is, however, the chondrophore, the spoon-like projection from the left valve at the hinge. Color is chalky white, although usually there are bits of dark periostracum clinging to the edge. A typical Soft-shelled Clam shell would not measure over 3 inches, though some of these shells may reach a length of 6 inches.

TRUNCATED SOFT-SHELLED CLAM, *Mya truncata,* taken from a depth of 15 feet at Henry Island, San Juan Islands, Washington. Note the hinge-supporting chondrophore.

TRUNCATED SOFT-SHELLED CLAM
Mya truncata Linne

The Truncated Soft-shell, or Truncated Mya, is a stubby clam reportedly found from Oregon into the polar region. The author has occasionally found them in the San Juan Islands of Washington, generally in shallow water, often not very far beneath the surface of the gravel and silt. Under water they can be gathered by grabbing the protruding neck and digging the clam out with a diving knife.

The shape of the Truncated Soft-shell starts out somewhat like the Eastern Soft-shell but the siphonal end is truncated, or chopped off. The shell alone, without the animal, resembles a little Gooeyduck. However, on the living clam a strong skin of periostracum starts along the edge of the shell and continues up the neck of the clam. When the clam is dug and the siphon retracted, this heavy skin has a wrinkled, dirty-brown appearance, so the clam at times is discarded. It is, however, a satisfactory eating clam in stews and chowders.

Shells of the Truncated Soft-shell Clam are readily distinguishable by the exterior texturing of the heavy growth lines, the deep pallial sinus inside the shell, and the spoon-like chondrophore. The two valves are of inequal size. Length is seldom over 3 inches.

GIANT CLAMS

Strange as it seems, size is not a common characteristic of Giant Clams. Most Northwest members of this family are small, uninteresting clams that are found boring into sponge and soft rock. Common characteristics are the irregular and rough shells (at times not of equal size) and, on the living animal, valves that gape so widely they cannot be closed. *Hiatellidae*

GOOEYDUCK (GEODUCK)

Panope generosa Gould

The largest clam of the Pacific Northwest, and the species most popular with clam diggers in the Puget Sound and Hood Canal regions of Washington, is the Gooeyduck. It has one of the largest shells (up to 8 inches in length) and is the largest clam by body weight of any found in this area. Digging the Gooeyduck is a challenge, for this clam, with its 3-foot siphon, lies deep in the mud and gravel of the tide flat. Underwater harvesting of the Gooeyduck has been attempted, using the water jet from a high-pressure hose to dislodge the clams, but reportedly this has not been very profitable. The meat of the Gooeyduck is presently marketed as "King Clam."

The shell of the Gooeyduck is oblong, rounded at the foot end and blunted or truncated at the siphonal end. Instead of the usual cupped shape, the shell is curved almost like a bent piece of sheet metal. Not only do the shells gape widely at the posterior end (where the siphon is) and the anterior end, but they fail by an inch or more to meet at the ventral portion as well. The two valves, in fact, seem not to be matched. Such gaping, particularly at the anterior end (the foot), is rather typical of the boring clams to which the Gooeyduck is related.

The texture of the shell consists of rough, concentric growth lines. The interior below the hinge and above the pallial line is also roughened, but not in a pattern like the exterior. At times the interior roughening includes ridges angling toward the center of the shell. Color is a chalky, grayish white. Gooeyducks are normally found with a partial covering of yellowed periostracum.

GOOEYDUCK, *Panope generosa,* dug from the tide flats near the mouth of the Dosewallips River, Washington.

SURF CLAMS

As the name implies, most Surf Clams live in sandy conditions and often in the surf of the open coast, but not so our Pacific Northwest members of this family. While ours have the same equivalve characteristics with the tendency of the shells to gape at one end, they are found in silty bays, estuaries, and sheltered waterways. Only one, the Gaper Clam, is sufficiently abundant to be used for food. *Mactridae*

GAPER CLAM

Tresus capax (Gould)

The Gaper Clam shell competes with the Gooeyduck in size, but the body is smaller. Actually, the names of our two largest clams could be exchanged, for compared to the Gooeyduck, the Gaper shells gape very little away from the siphonal area. There, the shells are rounded out to allow the large siphon always to appear, even when the "neck" is withdrawn. Under water, the extended siphons of these two large clams seem about the same size, but the siphon of the Gaper is covered with a thick, black skin and has small flaps at the end.

The Gaper is not as highly prized for its eating quality, but does not dwell as deeply under the mud and gravel so is more readily obtained, and it is much more abundant. Its range is also larger, extending well into Oregon, while the Gooeyduck is rarely encountered in Oregon, and then only in one location, Netarts Bay. Another species of the Gaper Clam is *Tresus nuttalli,* which is fairly common in Washington but rare in Oregon.

Shells of the Gaper Clam may reach 8 inches in length. They are a chalky white on the exterior, though somewhat shiny on the inside. The edges normally are covered with the remnants of a dark brown periostracum that forms like a skin as the shell grows, but later wears and peels off. The shape is unique, sloping with moderate steepness on each side of the beak. Because the posterior or siphonal slope is longer, that end of the clam seems to sag. The foot end is shorter and rounder. On the outside of the shell are roughly indented concentric growth rings. The interior is smooth but shows a very large pallial sinus. An interior characteristic of this and other Surf Clams is a large pit where the hinge ligament attaches.

GAPER CLAM, *Tresus capax,* dug from the tide flats at Yaquina Bay, Oregon.

HOOKED SURF CLAM, *Spisula falcata,* found at a depth of 40 feet in Gamble Bay, Washington.

HOOKED SURF CLAM
Spisula falcata (Gould)

The Hooked Surf Clam is occasionally found from British Columbia southward, both on the open coast and in sheltered waterways. Because it is not common anywhere in the Pacific Northwest, it is rarely sought for food.

Superficially, the shape of the Hooked Surf Clam looks like its larger relative, the Gaper Clam. Both shells slope away from the beak; one end is longer and narrower than the other. However, when the inside portions are compared, it is evident that it is the anterior (foot) end of the Surf Clam that is longer and narrower, not the siphonal end like the Gaper. Also, the valves of the Surf Clam do not gape; they are capable of closing tightly. A common characteristic of both members of the Surf Clam family, though, is the hinge. Both clams have rounded pits into which the hinge ligaments attach, and both have hinge teeth that interlock the shells.

Shell color of the Hooked Surf Clam is brownish white, but shells of the living creature are covered with a darker brown periostracum. On this periostracum, growth lines are readily visible in the concentric streaks of light and dark brown. Shell length of adult Hooked Surf Clams is about 3 inches.

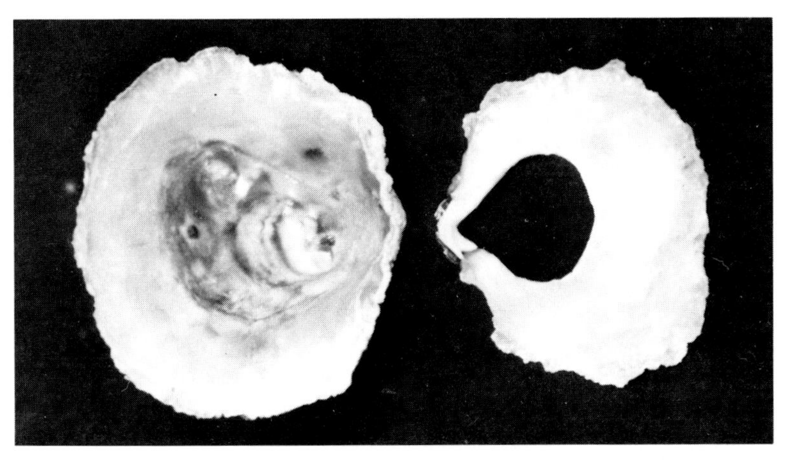

PEARLY MONIA, *Pododesmus cepio,* collected from the south jetty of Yaquina Bay, Oregon. The pearly interior of both valves is shown, with the upper valve on the left.

JINGLE SHELLS

Members of this family have thin, translucent, and inequal valves. They attach to a hard object by a fleshy byssus which passes through a hole in what becomes a lower valve, remaining stationary there through life. Most are residents of warmer seas. *Anomiidae*

PEARLY MONIA

Pododesmus cepio (Gray)

Pearly Monias can frequently be found attached to rocks, shells, and other hard objects in quiet coves, jetties, and sheltered waterways. The lower valve, through which the byssus passes, is thin and conforms tightly to the surface of whatever it attaches to. The upper valve also follows this configuration, making the Pearly Monia both difficult to see and extremely hard to remove without breaking the fragile shells.

The shell exterior is normally roughened and brownish. The interior, however, is pearly and iridescent, usually blushed with green. (A similar species, *P. macroschisma,* which is dominant in northern Washington and British Columbia, usually has brown tones.) Size may be as much as 4 inches across; the lower valve, with the teardrop-shaped hole for the byssus, is slightly smaller.

PIDDOCKS

Clams in this family have an abrading texture at the anterior (foot) end with which they bore into fairly hard surfaces like wood, concrete, and some rock. Shells taper toward the siphonal end, as do the holes into which they bore. Piddocks are found world-wide. *Pholadidae*

ROUGH PIDDOCK

Zirfaea pilsbryi Lowe

The Rough Piddock is the largest member of this family in the Pacific Northwest, at times reaching 4 inches in length. It lives in shallow water, often in estuaries, where it bores into hard clay or soft rock.

Shells of the Rough Piddock are divided into two sections by a belt-like groove. The siphonal end looks like an elongated clam whose white shell tapers to a rounded and gaping tip. Texture on this half consists of concentric growth lines. The anterior half, however, is triangular, leaving a huge opening for the foot. This part is textured with frequent angled growth lines crowned by radiating rows of saw teeth. This is the rasping instrument with which the Rough Piddock bores. A myophore, shaped like a curved grooving chisel, lies below the hinge in each shell. This is a muscle attachment. Also near the hinge is a fold of shell material that extends out, hiding the beak.

COMMON PIDDOCK

Penitella penita (Conrad)

The smaller Common Piddock is widely distributed along the open coast where it bores into any moderately hard rock. It is harvested for food by breaking apart the rocks with a hammer. Living in the intertidal zone, it is more accessible than the Rough Piddock.

Shells of the Common Piddock are somewhat like those of the Rough Piddock; the siphonal end is elongated and tapering, marked by growth lines, while the anterior has triangular markings with saw-like teeth. However, the foot is completely covered by a smooth bulbous area of shell. Its texture and fragility lead to some speculation that this portion may be temporary, lost during active boring, then re-secreted. Length may reach 3 inches.

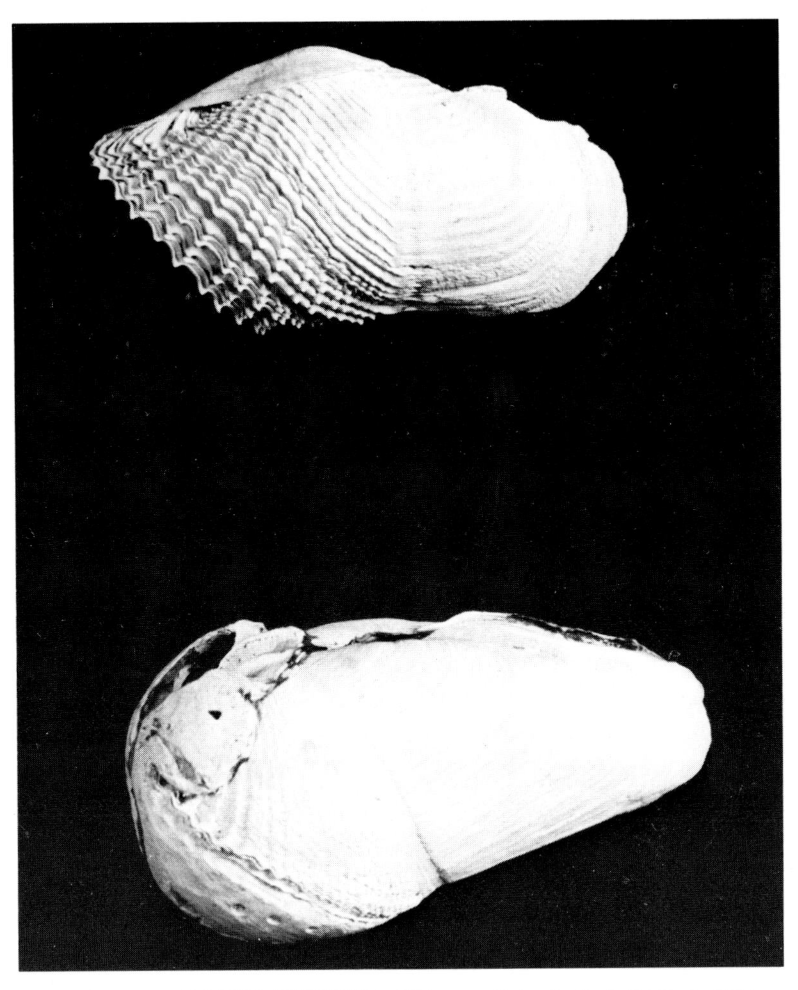

Top: ROUGH PIDDOCK, *Zirfaea pilsbryi,* dug from a clay bank under 30 feet of water in Yaquina Bay, Oregon.
Bottom: COMMON PIDDOCK, *Penitella penita,* chiseled from intertidal rocks near Seal Rock, Oregon.

OYSTERS

Bivalves in this family lose their power of locomotion in the adult stage, for they attach to a hard surface and remain there for life. Lacking the extendible siphons common to clams, they are suffocated if their site becomes covered by debris. Northwest species of Oysters attach by cementing the lower valve to the attaching site. *Ostreidae*

NATIVE OYSTER

Ostrea lurida Carpenter

Also known locally by such names as Olympia Oyster and Yaquina Oyster, this bivalve is small (normally around 2 inches in length) but tasty. Original populations, through their exportation, played important roles in the development of several Pacific Northwest communities. They are still gathered for food and are now cultivated commercially.

Shells of the Native Oyster are shaped into various forms of distorted ovals. The lower (fixed) valve is cupped or concave, but the upper valve may be nearly flat. Texture is generally absent although lumps and growth rings occasionally appear. Edges sometimes seem flaky. Shell color is dull white with blotches of gray that give it a burned appearance. The interior has partial iridescence, but this is not an attractive shell.

PACIFIC OYSTER

Crassostrea gigas (Thunberg)

The Northwest's largest oyster, the Pacific, was originally imported from Japan. Seed oysters continue to be brought in for commercial beds, but in the sheltered waterways of Washington and British Columbia they reproduce naturally, spreading to other rocky or gravelly intertidal areas.

The shape of Pacific Oysters varies, but tends toward a long oval, pointed at the hinge. Mature shells develop a wavy or fluted appearance at the edges, which is a distinguishing characteristic. They are dull white, at times with black on the fluting. Length may reach up to 12 inches.

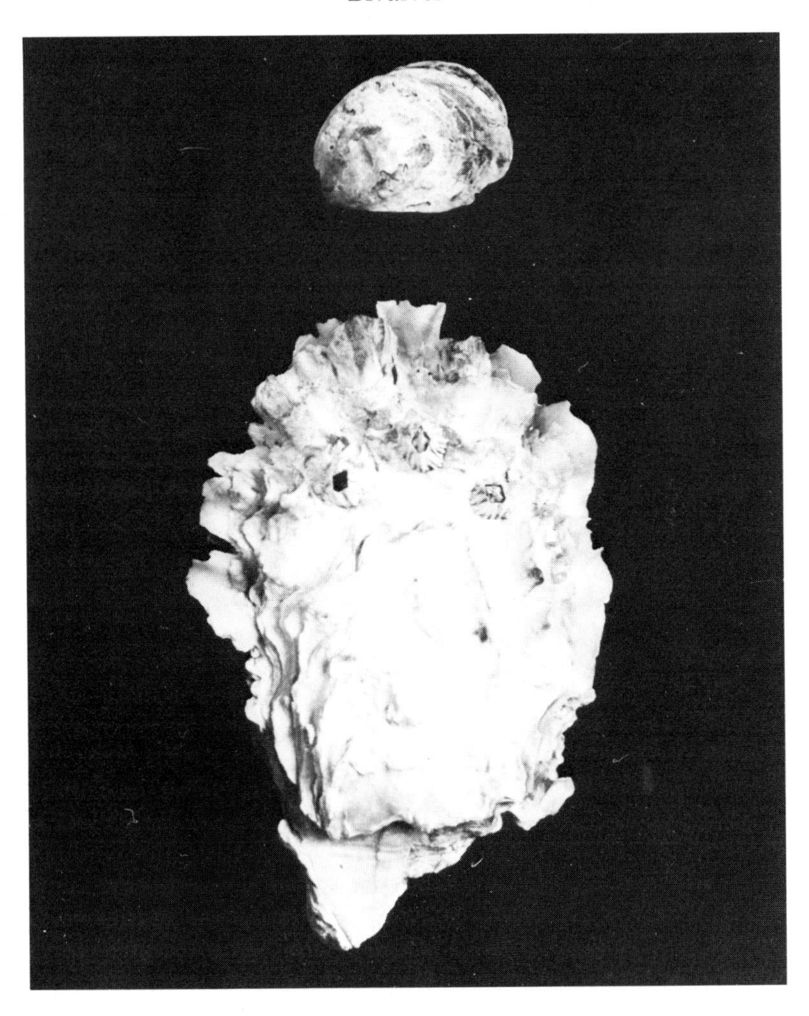

Top: NATIVE OYSTER, *Ostrea lurida,* removed from intertidal rocks in Yaquina Bay, Oregon.
Bottom: PACIFIC OYSTER, *Crassostrea gigas,* gathered intertidally at Seabeck, Washington.

MUSSELS

Bivalves of this group lack extensible "necks" or siphons. They retain limited mobility throughout their lives but often attach to hard surfaces by means of a byssal strand they manufacture and attach with their foot. The interior surfaces of their shells are pearly, the exterior covered with dark, heavy periostracum. Mussels are edible and frequently used as food, particularly in other lands. *Mytilidae*

SEA MUSSEL

Mytilus californianus Conrad

Beds of Sea Mussels cover vast areas of intertidal rocks along the open coast. In the beds they are normally found so close together that they stand on end, attached to the rocks and to each other by a number of byssal threads that the animal readily manufactures. To some extent, the Mussel has limited mobility with these strands, having been observed in aquariums to use them occasionally to pull themselves to a higher position.

Sea Mussels are used for food at times but in water conditions warmed by the summer sun they may accumulate toxins from then-abundant organisms which they filter from the water and ingest. Thus they are not regarded as edible from May through August.

The beak of the Sea Mussel shell is located at one end, while the opposite end is rounded, giving it a shape like a small boat. The ventral edge is shorter and straight, however, so the boat is lopsided. In life, the two valves are connected by a long, rather thin hinge ligament that extends almost halfway back from the pointed end. This makes the shells, which fit tightly together, extremely difficult for predators to open. Texture, which is a differentiating characteristic of the Sea Mussel, consists of roughly indented lines of growth and about six or eight lengthwise grooves on adult specimens.

A dark blue, almost black, heavy periostracum covers that portion of the shell where it has not been worn away. The shell material also is tinted a dark blue. The usual length of Sea Mussel shells in intertidal beds is around 3 inches. However, some areas with exceptional food potential may produce Sea Mussels 8 inches in length.

The SEA MUSSEL, *Mytilus californianus,* on the left, was collected from intertidal rocks at Lincoln City, Oregon. Note the lengthwise grooving, compared to the smooth shell of the BAY MUSSEL, *Mytilus edulis,* collected from piling in Yaquina Bay, Newport, Oregon.

BAY MUSSEL
Mytilus edulis Linne

While the Sea Mussel is found on the open coast and occasionally among the islands of Washington and British Columbia, the Bay Mussel is virtually confined to bays and quiet waterways. There it is often found in abundance but not in dense beds like the Sea Mussel. It is, however, not usually in the same environment as the organisms that produce paralytic shellfish poisoning, so is considered safer for human consumption.

Shells of the Bay Mussel are white with blue tinting, covered with dark blue periostracum (juvenile mussels may be brown). Bay Mussels are not as boat-shaped as Sea Mussels because of their definite curvature. Distinguishing characteristics are the inward curve of the ventral margin, the curve, or hump, of the dorsal margin, and the smooth texture. Size is usually less than 3 inches.

GIANT HORSEMUSSEL

Modiolus flabellatus (Gould)

Unlike the mussels of the genus Mytilus, the Giant Horsemussel lives in the mud of sheltered waterways and deep water offshore. It also differs by being solitary, rarely living in colonies. Since most of them are found at depths of over 20 feet, their shells are seldom picked up by beachcombers.

The unique shape of the Horsemussel shell somewhat resembles the human foot. The slightly curved ventral line is shaped like the inner edge of a foot; the tapered and sharply rounded anterior end is like the heel, and the dorsal edge is like the outer profile of the foot, widest at the little toe, with the posterior end sloping like the toes. Even the beak is well anterior, positioned about where the ankle bone would appear, with a ridge running from the beak to the "big toe" like the top of one's arch.

Giant Horsemussel shells are scaly looking, dull white with peeling, dark brown periostracum. Inside, the shells are iridescent white, often shaded with pink, and at times studded with tiny pearls. Length may reach 7 inches.

NORTHERN HORSEMUSSEL

Modiolus modiolus (Linne)

The Northern Horsemussel is a smaller mussel, about 3 inches in length, and is found both in the North Pacific and the North Atlantic. There it lives in rocky and muddy bottom areas at subtidal depths where it is occasionally discovered by skin divers.

The shell of the Northern Horsemussel is smooth, moderately elongated and boat-shaped. Its beak is almost to the anterior end. A hinge ligament runs nearly a third of the length of the humped dorsal edge. The siphonal end of the shell is rounded and the anterior has a rounded point. The Northern somewhat resembles the Giant Horsemussel but is less "foot-shaped" and lacks the broad ridge angling across the shell. Shell color is white, a little pearly inside, with a smooth brown periostracum covering the exterior.

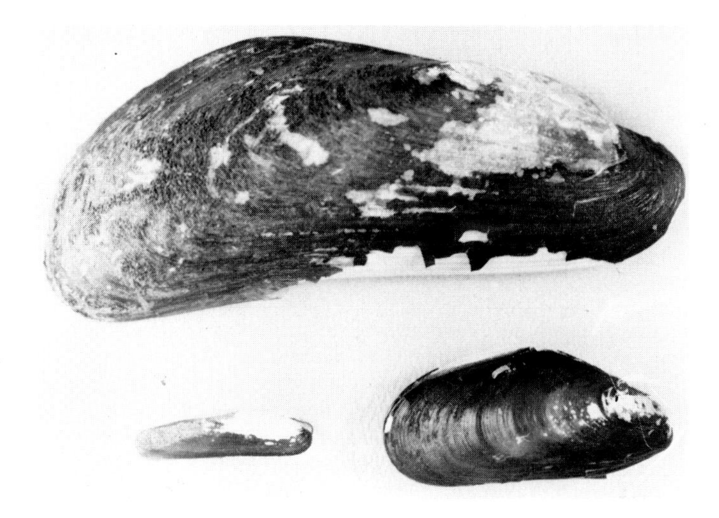

Top: GIANT HORSEMUSSEL, *Modiolus flabellatus,* collected from Hood Canal near Brinnon, Washington.
Bottom: CALIFORNIA PEA-POD SHELL, *Adula californiensis,* left, pried from intertidal rocks near Seal Rock, Oregon; NORTHERN HORSEMUSSEL, *Modiolus modiolus,* above, found in the same subtidal area of Hood Canal as the Giant Horsemussel.

CALIFORNIA PEA-POD SHELL
Adula californiensis (Philippi)

The California Pea-pod Shell is the most common and largest of several small, slender mussels that burrow in clay and soft rock. It is found intertidally and in shallow water along rocky parts of the coast.

The shell of the California Pea-pod, as the name implies, is long, slender, and small, usually not over an inch and a half in length. Both ends are rounded, while the ventral and dorsal edges are nearly straight. Because the Pea-pod is fairly thick-bodied, the shells are full or deeply concave. The appearance of the living clam is rod-shaped, although the shells flatten toward the siphonal end, somewhat wedge-like. Color is white beneath a dark brown periostracum. When the periostracum is worn away, the thin, fragile shells seem almost transparent.

Collecting and Preserving Shells

Serious shell collectors prefer gathering live-taken specimens. Taking the shell before its occupant has died, however, does not endanger the population if one is careful to select only perfect examples, leaving immature creatures to grow up and reproduce and allowing those with damaged shells to remain in the active population. The conscientious collector will also make sure that the shells he takes are not wasted, that care is taken in cleaning and preserving, and that the shells are displayed in a manner to protect their beauty.

The inexperienced shell gatherer may be inclined to pick up too many and not be selective. Faced with the task of cleaning out odorous contents, he may discard everything. In this way, whole areas have become denuded of their shellfish beauty, and strict conservation laws have become necessary.

There are several problems in cleaning and preserving specimens. The most pressing one is that of removing the body of the animal. Upon death, organisms decay, producing a strong, offensive odor. This is a special problem for collectors while traveling, and has resulted in tales of unfriendly looks or caustic comments as the collector hurries along with the now-identified odorous flight bag.

One method of preventing such an occurrence is by soaking the specimens in alcohol and removing them just before packing for travel. This will delay the development of odor-producing bacteria. Formalin, which hardens the tissue, should be avoided because it makes the tissue difficult to remove later.

There are several ways of removing the flesh from snail shells. Some prefer repeated freezing and thawing to loosen it, then hooking the meat with a sharp instrument like a nutpick and pulling it out. Others place the snails in a pan of water and bring it to a boil. After simmering the animal for a few minutes to a half hour—depending upon the size—it can be cooled and the flesh removed in the same way. Care should be taken in cleaning not to change the temperature of the shells too rapidly as this may cause cracking. With snails, the operculum, or trapdoor, should be saved, dried, and glued to a piece of cotton. When pushed back into the shell the operculum is again in place.

Cleaning the exterior of a shell often requires considerable scrubbing. An old toothbrush, soap, and water are the basic materials. Much dirt and algae can be removed by soaking the shells in a solution of bleach and/or detergent. Abalone shells, though, should not be subjected to laundry bleach, as this will oxidize the surface and fade the colors. Finally, the color of the shells can be restored to its original glistening, damp appearance by rubbing with light mineral oil. Thus taken care of, shells will keep their "touch of the sea."

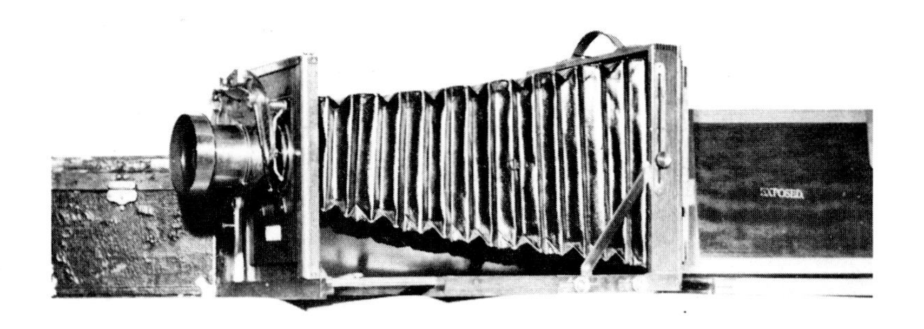

Photographic Techniques

Photographic techniques used in the illustration of this book varied widely, from the use of a modern waterproof camera with a self-metering underwater strobelight to the return from retirement of the large view camera once used by scientists of earlier days. Color plates, 4x5 Ektachrome transparencies, of the Frilled Dogwinkles (page 14), Smaller Dogwinkles (15), Japanese Oyster Drills (22), Pink and Hinds' Scallops (74), Cockles (87), and Thin-shelled Littleneck and Lucine Clams (95) were taken with an Athlete view camera (Pat. 1886), made by the Rochester Optical Company, and previously used by pioneer botanist, Francis M. Fultz. The plates are reproduced here in actual size. Other transparencies of preserved shells were taken in 2-1/4x2-1/4 Ektachrome, principally with a Kowa-Six single-lens reflex camera. All such transparencies were taken in natural light, using white paper and, at times, a mirror to diffuse shadows.

Many of the black and white pictures used were taken with a Praktica single-lens reflex camera in 35mm, with close-up views achieved through a Vivitar variable close-up attachment. These include the groups of Frilled and File Dogwinkles (page 13), Drilled Shells (21), Whelk and Hairy Snail (25), Wentletraps (29), Tabled Neptune (57), Smooth Washington Clam (89), Kennerly's Venus (91), Pacific Razor Clam (97), Gooeyduck (103), Gaper (105), Pearly Monia (107), and the Mussels (113 and 115). The remaining black and white photographs were taken with the Kowa-Six. Tri-X Pan film and natural light were used, and shutter speed adjusted to give f22 lense opening.

Underwater photographs of the Pinto Abalone (title page) and Moon Snail (43 and 46) were taken with a Nikonos, using 35mm High-Speed Ektachrome and a Honeywell Auto Strobonar 772 for light source. Intertidal photos were taken with the Praktica, using natural light.

List of Illustrations

Glossary

Anterior: The end that moves forward—the head in chitons and snails, and the foot end in bivalves.

Aperture: The opening of the shell through which the animal extends and withdraws.

Apex: The tip of the spire, often containing the nuclear whorl, from which the shell was built.

Beak: That portion of other valves (shells), corresponding to the apex of a snail shell, from which growth extends. Normally it is raised and tapered like a bird's beak.

Bivalve: A mollusk with two shells or valves, normally hinged together as two nearly equal halves of the protective cover.

Body Whorl: The newest and largest whorl (turn) of the shell of a gastropod, terminating in the aperture and containing the major portion of the body.

Byssus: A fibrous filament secreted by certain bivalves and used to attach it to hard surfaces.

Callus: A layer or deposit of shelly material, usually smooth and enamel-like, extending from the aperture and usually upon the body whorl, found in gastropods.

Canal: A tube or groove-like portion of the aperture of certain gastropods, at the nose; it may be open (groove-like) or closed.

Chondrophore: A projection on the inside of some bivalve shells at the hinge, shaped like the blade of a spoon.

Columella: The center-line around which the gastropod shell spirals as it grows.

Concentric: Encircling at intervals, as the waves from where a pebble is dropped; in bivalves, the beak is at the center.

Dorsal: Pertaining to the back. With bivalves, that edge from which the shell grows—usually containing the hinge.

Ear: A flat protrusion to the side (or sides) of the beak of a bivalve, most common on scallops.

Flaring: Spreading outward, usually at the edge (or previous edge) of a mollusk shell.

Foot: A muscular organ used for locomotion by most mollusks.

Genus: The first-listed and capitalized portion of the scientific name; pertaining to the group of species with similar characteristics.

Girdle: A tough, gristly band of material at the edges of the plates of chitons, often decorated with short bristles or scales.

Growth Lines: Lines left in shell surfaces by periods of slowed growth, usually shaped like a previous lip in a gastropod shell, like the rings of a tree on bivalve shells.

Head Valve: That one of the eight plates of a chiton that is closest to the "head,"—or that end of the animal that contains the mouth and sense organs.

Hinge: The place where the two valves of a bivalve are joined together by the ligament.

Inner Lip: The edge of the aperture bordering the columella.

Intermediate Valves: The six plates or valves extending from the head valve to the tail valve of a chiton.

Ligament: Hinge material that is flexible in the living bivalve, forming a movable connection that allows the valves to spread.

Lunule: A heart-shaped pattern anterior to the beaks on some bivalves—one half on each valve.

Mantle: The surface portion of the body parts of mollusks, external and tough in some cases, bordering and secreting the shell in others.

Myophore: A peg-like projection inside a bivalve shell used to support a ligament or muscle.

Nose: The extended part of the aperture of a gastropod; opposite end of the shell from the apex.

Nuclear Whorl: The initial portion of a snail shell, generally present when the snail hatches, and often smooth in texture.

Operculum: The trapdoor possessed by most marine snails that seals the aperture when the snail withdraws into the shell.

Outer Lip: The outer edge of the aperture, primary growing part of the shell.

Pallial Line: A line around the inside edge of a bivalve shell marking the attaching line of the thick edge of the mantle.

Pallial Sinus: An indentation in the pallial line near the posterior end where the siphon lies in the shell.
Periostracum: A skin-like material that covers many shells.
Plates: Overlapping pieces of hard, shelly material that form the body armor of the chitons.
Posterior: The trailing end in mollusk movement; in chitons and snails, the opposite end from the mouth; in bivalves, the end with the siphons.
Proboscis: Tube-like mouth of a predatory snail used to reach into its prey.

Radiating: Reaching from the beak or apex of a shell toward the distant edges; as opposed to concentric.
Radula: A mollusk's tongue. It contains rows of rasp-like teeth.

Siphon: A tube-like organ which is used to draw water into the mantle cavity for food-filtering and/or respiration, or to exhaust water. It may be fused together as a clam neck, separate tubes in some bivalves, or a single fold of tissue in snails.
Species: The second part of the scientific name denoting the particular group to which an individual animal belongs.
Spire: The earlier whorls of a snail, exclusive of the body whorl.
Suture: The line formed where gastropod whorls join together.

Tail Valve: The end valve of the eight chiton plates—opposite the head valve and closest to the anus.
Tooth: A small protruberance in the aperture of a snail shell, most frequently appearing as a row of teeth on the outer lip, or as a single tooth on the inner (columella) surface; also a term used to denote hinge projections on bivalves.

Umbilicus: A gap or hole left between the aperture and the body whorl of some gastropod shells as the shell grows, most common on globular shells.

Valve: Unit of shell, two of which make up the shelly covering of a bivalve, eight of which make up the covering of a chiton.
Varix: A ridge crossing a gastropod shell whorl lengthwise to the shell.
Ventral: Pertaining to the stomach; the edge of a bivalve shell opposite the hinge and usually rounded.

Whorl: A spiraling turn or volution of gastropod shell growth.

Index to Scientific Names

O

P

S

T

V

Z

Index to Common Names